W9-BQW-064

ENVIRONMENTAL AESTHETICS:

Essays in Interpretation

ENVIRONMENTAL AESTHETICS:

Essays in Interpretation

edited by
BARRY SADLER and ALLEN CARLSON

Western Geographical Series Volume 20

Department of Geography, University of Victoria
Victoria, British Columbia
Canada

1982 University of Victoria

Western Geographical Series, Volume 20

editorial address

Harold D. Foster, Ph.D.
Department of Geography
University of Victoria
Victoria, British Columbia
Canada

Publication of the Western Geographical Series has been gener-
ously supported by the Leon and Thea Koerner Foundation, the
Social Science Federation of Canada, the National Centre for
Atmospheric Research, the International Geographical Union Con-
gress, the University of Victoria and the Natural Sciences and
Engineering Research Council of Canada.

Copyright 1982, University of Victoria

ENVIRONMENTAL AESTHETICS
(Western geographical series; ISSN 0315-2022; v. 20)
"Papers . . . originally presented at a symposium entitled the Visual
Quality of the Environment held at the University of Alberta in
September 1978."
ISBN 0-919838-10-3

1. Landscape assessment — Addresses, essays, lectures. 2. Nature
(Aesthetics) — Addresses, esays, lectures. I. Sadler, Barry. II.
Carlson, Allen A. III. University of Victoria (B.C.), Dept. of Geog-
raphy. IV. Visual Quality of the Environment (1978: University of
Alberta) V. Series.

GF90.E59   304.2   C82-091058-9

ALL RIGHTS RESERVED

This book is protected by copyright.
No part of it may be duplicated or reproduced
in any manner without written permission.

ACKNOWLEDGEMENTS

This book has been published with the help of a grant from the Social Science Federation of Canada, using funds provided by the Social Sciences and Humanities Research Council of Canada. The two editors of the volume also wish to acknowledge the financial support provided by the Environment Council of Alberta and the University of Alberta for the symposium at which the original drafts of these papers were presented.

Many members of the Department of Geography of the University of Victoria have assisted in the production of this volume. Graphic work was undertaken by staff cartographers Ken Quan and Ken Josephson. Diane Brazier undertook the demanding tasks of both typesetting and layout. The assistance of these individuals is very gratefully acknowledged.

University of Victoria   Harold D. Foster
Victoria, British Columbia   Series Editor
Canada

April, 1982

PREFACE

Environmental aesthetics is now the subject of serious study by geographers. Such a statement, in itself, summarizes the changes that have occurred in thought and research since 1972, when a small panel of distinguished scholars met to consider the specific problem of *Visual Blight in America* and issued a call for action on the broader front. A steady expansion has taken place in the volume of writing that now deals, in one way or another, with the aesthetic qualities of place and landscape. It is linked to the development of broader frameworks of geographical enquiry, which reflect a renewed interest in human experience and social values.

This emphasis has moved geographers towards closer ties with philosophy and the humanities, much as the earlier behavioural revolution created links with psychology and the related sciences. Such inter-disciplinary collaboration holds some promise for the development of more holistic and eventually integrated prespectives on man's aesthetic relationships with the environment. We have tried to make a modest but formal beginning to this process in the present volume. It brings together contributions to the field from scholars drawn from philosophy, literature and landscape architecture, as well as geography.

The papers were originally presented at a Symposium entitled *The Visual Quality of the Environment* held at the University of Alberta in September 1978. At that meeting, the interaction between the participants and the response of the audience, encouraged us to commission revised papers for publication. Our objective in preparing this special volume is to present a suggestive review of "the state of the art" of environmental aesthetics that may promote further dialogue and discussion. It is designed to exemplify the main themes and illustrate the diversity of approaches found in the field. More idealistically, perhaps, we hope that the volume might stimulate a more widespread concern for the aesthetic quality of the environment.

The reasons why our optimism must be tempered with realism is illustrated by a minor issue which resulted in a certain degree of local interest and media coverage of the Edmonton symposium. It is an unremarkable story of transient public reaction aroused by a small-scale incidental change of the kind that is probably duplicated somewhere in North America almost every day. In this instance, it was the destruction of the continuity of a long line of mature trees to accommodate a minor road alignment. The action attracted notice because this particular shelter belt is a roadside landmark for regular travellers on the Edmonton-Calgary highway, one that establishes a sense of location and duration of journey across the relatively undifferentiated topography of the Alberta High Plains. A brief period of muted anger on the part of a small minority soon died away. It is questionable whether the concern even reached the attention of the highway engineers responsible, certainly the criticisms of their action went unanswered. The point made here is that much of the decline in the aesthetic quality of the environment is typically piece-meal and insidious, the by-product of a multitude of planning and development decisions taken with other criteria in mind. Such changes go either unnoticed or unchallenged although the cumulative affect on landscape is considerable. To the extent that the papers in this collection yield a better appreciation of the aesthetic value of the environment and their susceptability to impairment they will have served a practical as well as an academic purpose.

Barry Sadler   Allen Carlson
Victoria, B.C.   Edmonton, Alberta
April, 1982

PLATE 1 The Environmental Display.
Photos by B.C. Government except lower left by Vancouver Sun ▶

TABLE OF CONTENTS

LIST OF FIGURES

Figures

LIST OF PLATES

Plates

PLATE 2 Environment as Art: The Classical Legacy — Roman Aqueduct at Tarragona.
B. Sadler Photo ▶

xiv

1 ENVIRONMENTAL AESTHETICS IN INTERDISCIPLINARY PERSPECTIVE

Barry Sadler
University of Victoria
and
Allen Carlson
University of Alberta

INTRODUCTION

On a human scale, the environment is a continuing theatre for sensing and acting. At every moment, Kevin Lynch has noted, "there is more than the eye can see, more than the ear can hear".[1] Environmental perception thus is inevitably selective and organised by purpose. It is most often directed towards the cues and information necessary to perform certain everyday tasks. Few people, however, remain impervious to the quality of their surroundings for very long. The drive to work along urban expressways, for example, becomes a visual experience in which motion, space and form are interconnected and synthesised.[2] We all interpret, in one way or another, the positive and negative features of the physical settings we occupy, pass through and visit.

The aesthetic effect of places and landscapes is an important dimension of this pervasive sensory ecology. A sense of beauty or even harmony enhances our lives; a sense of blight or discordance correspondingly diminishes them. It is almost impossible to prove scientifically these kinds of effects on well-being, but the general principle is widely accepted. The President's Council on Recreation and Natural Beauty, for example, recorded this conviction in a report which now stands as a benchmark of official concern about the quality of the environment in the United States.[3] Similar

1

ideas about the power of the aesthetic properties of nature have long permeated and influenced aspects of thought and action in the conservation movement.[4] They also are expressed and endorsed in urban and regional planning. One of the important contemporary influences in this area has been the writings of Ian McHarg. His best known work contains a striking testament to the effects of environment on health and equanimity.[5]

At the same time, making aesthetic judgements brings into play certain well known problems. The interpretation of the value of beauty or the price of its loss is a notoriously subjective and slippery matter. Our standards of aesthetic quality for the environment are imprecise and indiscriminate. Except within very generalised limits, the conditions of scenic beauty resist specification. It is easier to define visual blight in the landscape, but this does not necessarily take us very far in terms of cultivating quality.[6] Nor is the form and process of the contemporary erosion of aesthetic values widely understood.

Visual eyesores, excessive noise and offensive smells are common and easily recognisable features of the urban and industrial areas where the majority of us now live. They provoke uniform disapproval. However, the nature and scope of aesthetic deterioration in the environment is both more widespread and more subtle than this. It involves nothing less than the disintegration of landscape and is grounded in the political and economic forces that shape growth in a modern, technological society. Earlier differentiation of the settlement pattern, expressed by coherent city, rural and wild landscapes, is being blurred by the amorphous sprawl of quasi-urban developments, the main visual characteristics of which are formlessness and sameness.[7] The fundamental changes in the appearance of the environment are thus occurring on regional and national scales and paradoxically are less detectable than sensory irritants at the localised level.

For this reason, geographers and other specialists in man-environment studies have been encouraged to become landscape critics.[8] The role entails a rediscovery and reworking of certain traditional areas of interest, crystallised by such concepts as *sense of place* and *regional character*, to determine the properties and values that constitute aesthetic quality in environmental contexts. It calls for the development of special abilities that combine sensibility to scenery and symbolism with knowledge of pattern and process.[9] One without the other runs the risk of either superficiality or sterility in aesthetic evaluation. To achieve the right blend, the professional skills of environmental and social scientists need to be brought into contact with the talents found in the arts and the humanities. In short, the task of criticism requires a genuinely interdisciplinary perspective, based on a synthesis of research and insight and related to the concerns of planning and design.

This requirement provides the rationale for the present volume. It represents the general point of departure for a wide-ranging survey of environmental aesthetics, one designed to establish and contrast the variety of themes and approaches that characterise the field of study. Within the compass of the volume, it is neither possible nor desirable to cover all the routes taken by work in environmental aesthetics. The emphasis instead will be placed on differentiating and interpreting certain cardinal directions. Such a review of the state of the art of the field seems necessary to lay the groundwork on which more comprehensive perspectives may be built.

The purpose of this introduction is to open the discussion by providing a textual review of the field. Our main concern is with emerging issues of theory and methodology and their implications for the practical contributions of research to environmental design and planning. It involves stepping back to selectively consider and exemplify some of the problems, as well as the possibilities, which accrue from the intermix of concepts and approaches of different disciplines. Studies in environmental aesthetics currently fall short of realising their potential because fundamental differences in philosophy and purpose restrict the generalisation of findings. A number of reviews of this and related questions have been conducted already in regard to specialised areas of investigation.[10] The next step involves trying to begin a broader process of integration. What follows is a prologue towards this end which may be helpful in setting the substantive papers in interdisciplinary perspective.

ON THE NATURE OF THE INQUIRY

Environmental aesthetics is an abbreviated term for an amorphous field of study. It is used for its scope of encompassing all or relevant parts of areas of research variously labelled as scenic assessment, landscape evaluation and urban imagery. All of these, in one way or another, deal directly with the intangible qualities of the environment which have some form of aesthetic appeal or significance for society. One of their shared characteristics revolves around the complexity and difficulty of identifying and interpreting the nature and scope of the values being sought.[11] This is the central problem in environmental aesthetics, a continuing question which all work touches upon or refers to in some form. More than any other aspect perhaps, it reveals the present stage and character of the field.

No organised body of theory or even common definitions are yet available. Such a lack of development is hardly surprising. The nature of the aesthetic has proved to be one of the most taxing and persistent issues in the history of ideas, while its application to the study of contemporary environments is a recent endeavour. Many studies, moreover, are concerned with developing and testing techniques for the evaluation of aesthetic quality.[12] It becomes important, in these circumstances, to examine the premises which guide research to ensure this is soundly directed. With the present theoretical vacuum, the student of the field, in effect, starts from "the same place as the man in the street".[13] At a minimum, we need to achieve some common understanding of the kind of questions raised by the nature of the inquiry. Our exploration of this topic begins with an analogy to art.

The concept of the aesthetic commonly is used to refer to the values inherent in works of art. A strict purist interpretation restricts the aesthetic to this realm, i.e. "to the deliberate creation of forms symbolic of human feeling".[14] It thus covers a specialised interaction between mind and form, one in which the trained eye or ear knows *what* to consider in works of art, whether painting, sculpture or music, and *how* to appreciate them. Different traditional schools of painting thus have different foci of aesthetic significance and demand different "acts of aspection." The critic, for example, pays attention to balanced masses in Venetian paintings and to contours in those of the Florentine school, and in doing so may survey one and scan the other.[15]

Under a broader interpretation, the features and patterns of the physical world also are recognised as having an aesthetic element.[16] The determination of this quality, however, is more problematic and open ended than is the case in the art world. Man's imprints on the landscape reflect not only functional necessities, but creative tastes and aspirations as well. Spatial arrangements and physical form, whether city streets or wild lands, record how communities have evaluated their surroundings and made them over partly in accordance with idealised images and visual stereotypes.[17] The physical expression of aesthetic values, however, is hard to decipher beyond certain generalised levels which are discussed later. It typically is ambivalent and ambiguous, permeated with other social, moral and ecological values, and usually represents only the predominant taste of an influential minority. For these reasons, there remains considerable uncertainty about the nature and degree of aesthetic intent in the humanised landscape, especially in relation to common buildings and popular styles.[18]

All environments, whether largely natural or extensively modified, nonetheless evoke feeling, and have some aesthetic dimensions. These aesthetic properties of the physical world include formal qualities of beauty, such as balance and contrast, which are highly valued in the arts and underlie much empirical research in environmental aesthetics, and a number of different kinds of non-formal ones. Expressive qualities, such as grandeur and serenity, are thought by certain commentators to be particularly important in environmental contexts. On contemplating a mountain lake, for example, "we see it not merely as an arrangement of pleasing colours, shapes, and volumes, but as expressive of many things in life, drenched with the fused associations of many scenes and emotions from memory and experience".[19] Urban centres also create a framework for the attachment of similar responses.[20] Environmental aesthetics, in other words, covers the storied meaning as well as the structured appearance of place and landscape.

The point being illustrated here is the loss of precision which occurs in transferring the concept of the aesthetic from the art world to the physical realm. As a result, studies collectively cover a diverse range of responses to environment. Settings and scales of investigation also vary widely. The attributes considered to be aesthetic, under these conditions, are more eclectic and emotively charged than those considered in "purist" judgements of art work. It thus becomes extremely difficult to categorise and compare the significant components of what is a particular mode of environmental experience. Some general system of classification is needed urgently to explore these relationships and guide selection in future research.

The difficulties of identifying the aesthetic qualities of the environment are product and cause of unresolved issues about how to study them. At present, two general avenues of approach are followed in the field. One leads towards the objective anchor of aesthetic qualities in physical reality; the other towards the subjective source of values which given them definition. The emphasis in the "objective" approach is on systematic analyses or quantitative measurements of the effect of physical variables (although the social and psychological factors that condition response are incorporated also to various degrees).[21] A recording of aesthetic values, from this perspective, may involve either a guild of experts or a random sample of the general public. In either case, the results are or should be verifiable or reproducable. By contrast, the "subjective" approach is grounded in humanism and places stress on the intrinsic meanings of the environment. The relevant values are revealed by personal introspection or studied reflection on the totality of experience, and here aesthetic response is considered to be fleeting and unexpected. It is not dependent on the particular character of an environment and is "the antithesis of the acquired taste for certain landscapes or the warm feelings for places that one knows well."[22] The "subjective" and "objective" approaches to environmental aesthetics, of course, are not mutually exclusive in practice, but overlap at a number of points.

It is useful to correlate these approaches with opposite poles of a continuum. Aesthetic quality, however defined, is an amalgam of physical properties and social values. Not only are these constituents hard to disentangle in any meaningful way, their combination fluctuates with the environmental contexts and social reactions under consideration. The methodological issues involved in determining the relative weightings and the factors which explain variance are complex and cannot be discussed here. The important point to note instead is that both "objective" and "subjective" positions are necessary for a complete and balanced account.

Whether these can be brought together effectively is another matter. One problem is the wide standards of evidence which are accepted in such a broad interdisciplinary field as environmental aesthetics.[23] Experimental work in stimulus arousal, the phenomenological interpretation of environmental gestalts, and historical studies of national attitudes are cases in point which range along the spectrum of approach. The possibilities for fruitful collaboration between those steeped in broad humanist and narrow empiricist traditions is especially open to conjecture. Even in specialised subfields, the basis on which interpretations are made is unclear or inconsistent. Some surveys of environmental perception, for example, distinguish between considered assessment of scenic quality and personal preference for landscape forms. Others make no such separation. The theoretical and practical implications of these distinctions are considerable and remain a source of concern in the field.[24]

Beyond the general recognition that certain landscapes are more visually attractive than others, no concensus exists on the relevant foci of aesthetic significance in the environment and the different acts of interpretation they demand. We have barely begun the task of building models applicable to understanding environmental aesthetics. Most empirical research tends to equate the aesthetic with the scenic and consider the power of effect as the measure of beauty. Other studies, notably the more humanistic, suggest the profile of environmental aesthetics is narrow and deep rather than broad and shallow. In the postscript to this volume, we return to this theme and attempt to sketch alternative paradigms of the aesthetic as it relates to the environment.

The immediate concern is to conceive the field as broadly as possible and include all relevant work. By general agreement, the aesthetic is yielded through a heightened phase of man's interaction with the environment and is an interdependent value being present in people when it exists in their surroundings and *vice versa*. Human response to the aesthetic aspects of the physical world range widely in content and depth from mere sensation to self-transcendence, from superficial pleasure in a passing view to a deep sense of oneness with the environment. As well as degree of intensity, another common denominator is the focus of perception. The environment, in all cases, becomes *appreciated* and revealed for its own sake. On this basis, we can proceed to develop guidelines for further analysis and extend these to consider the decline of aesthetic values.

6

A FRAMEWORK FOR FURTHER ANALYSIS

The environment as a source of aesthetic effect is one of vast scale, immense complexity and successive change. Sensory capacity and available technology set certain limits on the stimulus properties of the environment; even so, the potential content for aesthetic experience is seemingly endless. It encompasses the totality of our physical surrounds, from countless individual features to landscape as the assemblage of elements. All of this is another way of saying that it is important to consider and restate the basic questions discussed in the previous section in more concrete terms than is possible in relation to research theory and methodology. The emphasis in this section, in other words, is on developing firmer perspectives on what aspects of the environment are aesthetically appreciated and how people interpret them.

An initial framework for this purpose, based on the way physical settings are usually comprehended, is illustrated in Figure 1,1. The environment is organised into units of display which vary along two critical dimensions of scale and of character, expressed as the relative degree of natural or man-made influence. Figures 2,1 to 5,1 exemplify the aesthetic qualities of the elements included in the matrix. Our focus here is flexible and dependent on interest. It is important to stress, however, that some important distinctions are introduced as we move from the micro to the macro scale and from the natural to the man-made character of the environment. Noting these may help to clarify our understanding of the relationships between physical reality and aesthetic values and of the basic questions encountered in their analysis.

Small scale objects and groupings, best described as those which can be apprehended easily in their entirety, are reasonably analogous to works of art. Natural features and vistas are often appraised in this way and, in fact, constitute an important source of "raw material" for aesthetic creativity in the arts.[25] There also is a rich aesthetic tradition of design and criticism of individual buildings in the history of architecture.[26] However the very act of isolating elements for creative treatment or observation removes the most important and problematic aspect of the reality of environment and landscape, which is the integration of form in total context.

Yet large scale settings, such as natural regions or metropolitan areas, represent formats of a completely different order to the traditional objects of aesthetic scrutiny. This is a consequence partly of the extended scale of landscape perception, which occurs sequentially across space or intermittently over time. But also it reflects the multiform character of man's interaction with the environment. In the final analysis, aesthetic quality at the geographical scale can be comprehensively interpreted only in terms of the ecology of habitat.[27]

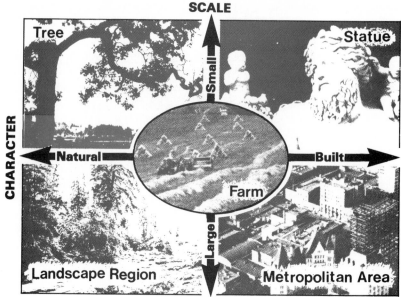

FIGURE 1,1   The Environmental Display.   after Craik, 1970

The important point to note in this context is the relatively minor role which aesthetic values play in the processes which govern the use of environment and the making of landscape. Only relatively few spatial enclaves are expressly designed or primarily set aside for aesthetic purposes. The great buildings and gardens of Europe are classic examples; their inherited traditions are reflected in the urban design on this continent (Figure 6,1). Aesthetic ideals were and continue to be instrumental in the establishment of national parks and similar reserves to protect the natural wonders and scenic landscapes of the American and Canadian west and subsequently of other parts of the world. Such examples, however, stand out as the exception rather than the rule, although most cities, for example, contain fragments of architectural merit. Landscape form generally is a vernacular product of the functional decisions of countless people, taken in circumstances and guided by traditions quite different from those involved in works of grand design or the establishment of parks.

On a macro scale, aesthetic quality must be understood largely in terms of the struggle for life and livelihood.[28] It is difficult to see how this could be otherwise in a world of hard choices and scarce resources. This is not to argue that "ordinary" landscapes lack aesthetic significance, only that this quality need be nothing more than the conscious and organised expression of man's encounter with the environment. Where aesthetic values are

8

FIGURE 2,1
Trees in the Forest.
These are Coast
Redwoods: their
aesthetic values
involve more than
symmetry and
texture in the play
of filtered light;
for they are among
the oldest as well
as the tallest living
things on earth.

FIGURE 3,1
Statue on the
Street.
The cheerful and
vulgar *Mannequin
Pis* is famous for its
associations not its
appearance. It is
steeped in legend
and rich in
commemoration for
the Bruxellois, and
is an obligatory stop
on tourist circuits
of the city.

FIGURE 4,1
Grand Canyon of
the Yellowstone:
Beauty and
Grandeur.
One of the natural
wonders which had
an aesthetic impact
on the region's early
explorers and led
to the birth of the
national park idea.

FIGURE 5,1
New York:
The archetypal
symbol of the
American City.
Its structure and
meaning is too
large to be grasped
in its entirety.
(In this case, the
skyline is exciting
and splendid
despite the non-
relationship of
elements).

dominant it may be appropriate to speak of the environment in terms of art, but in many cases we can look only for a sense of order and fitness, of form appropriate to function and to the character of site and community.[29] This gives a row of grain elevators punctuating prairie wheatfields or a well-knit inner-city neighbourhood an identity and relationship which is appealing to eye and mind (Figures 7,1 and 8,1).

The aesthetic quality of any environment, macro or micro, is open to a range of interpretations. No two people appreciate quite the same things even when looking at the same view. Our life biography, as well as mood and circumstance, condition what and how we sense the physical world. To a considerable degree though, aesthetic appreciation of our surroundings appears to be culturally consistent.[30] This is evident especially in the universal tendency to dichotomise and differentiate environments in relation to natural and human characteristics (i.e. along the horizontal axis of Figure 1,1). Aesthetic orientation in relation to the opposite poles of this continuum represents an important clustering and a stable organisation of values. This may be illustrated briefly by tracing the evolution and expression of certain concensual attitudes.

Nature has had an enduring impact on human emotion which reaches back to our beginnings and remains profound today.[31] Every culture and each age has its characteristic aesthetic view of nature and standards of beauty. A preference for mountains as scenery, for example, is widespread today among western society, although this is the end result of three centuries of changing appreciation of alpine regions, from the period in the seventeenth century when they were avoided as dangerous and abhorred as cataclysmic ruins of God's wrath.[32] Wilderness has undergone parallel change in the American mind: it is no longer seen as threatening or desolate, but as necessary and threatened.[33] The story of this change is closely linked to the history of the conservation movement and the fluctuating balance between utilitarian and aesthetic values. Modern environmentalist philosophy is based on bio-ethics rather than aesthetics *per se*: beauty is to be sought in the diversity and balance of ecological processes that are unimpaired by man and in the way of life that reveres and protects wild nature.[34] The natural condition of the landscape is a major factor also in determining popular appraisals of scenic beauty or visual appeal, although it remains open to question whether most people are able to discriminate gradations between the genuinely wild and the partly tamed.[35] Aesthetic sentiment for nature, however it is expressed, is often a mirror of anti-urban reaction to the ugliness of the modern metropolis.

Urbanism and civilization have been closely connected for millennia, although since the industrial revolution the idealisation of the city has been progressively eroded. The supremacy of the urban landscape as symbol of order and transcendence over wilderness has become reversed.[36] Now it is the city which is seen as a "chaos of non-relation",[37] where "the neglect of everyday life" is reflected in a pervasive blight of "fallen down buildings . . . grimy hospitals and bottle strewn parks."[38] The literary tradition of negative response towards the North American city is well known and this attitude is not confined to intellectuals by any means. For more and more people, however, the city and its regional extensions is the contemporary environment in which they act out their lives and is seen as having its own special aesthetic qualities and meanings.[39] Environmental appreciation of residential neighbourhoods typically is an incidental and unreflective aspect of attachment to locale which is manifest only under threat of change or after moving.[40] Aesthetic sentiment and symbolism are usually expressed for the public square, historic landmarks and civic monuments of the urban environment which allow us to make connections with the past as well as cement relationships with the present.[41] Urban centres thus carry an accumulated weight of positive as well as negative aesthetic values. In the final analysis, these tend to be more permeable and flexible than those related to nature so that it is possible to consider even Las Vegas as art form (Figure 9,1).[42]

Such observations within the context of Figure 1,1 may be helpful in organising an initial understanding of the foci and modes of aesthetic evaluation of the environment. But they tell us little about a more fundamental question: Why do we appreciate the physical world as we do? At present, this territory is a virtual *terra incognita* which only a few studies have attempted to chart. The experimental work of Berlyne on the power of particular stimuli to elicit arousal and curiosity provides a theoretical foundation for the psychology of environmental aesthetics. It is the basis for further studies of the correlations between affect and the collative properties of landscapes and places, such as complexity and novelty.[43] On this macro scale, however, Jay Appleton has suggested the origins of landscape appreciation are far deeper than those conventionally considered in research and lie in innate biological and behavioural traits.[44] His paper in this volume extends this provocative theory to develop a conceptual framework which permits a more fundamental understanding of environmental aesthetics.

PRACTICAL IMPERATIVES:
SOME SOURCES OF CONTEMPORARY CONCERN

A central dimension of work in the field of environmental aesthetics is a strong practical orientation.[45] Much of it is aimed at providing information which is useful to public policy decisions. The links between research and

FIGURE 6,1
The Great Tradition and the Little Tradition. The Pisan complex and the town beyond the walls illustrates the distinction between the formal and vernacular translation of taste into form, between conscious and unself-conscious aesthetic striving.

FIGURE 7,1
Grain Elevators on the Prairie. Like Barns, a fine example of instinctive vernacular design. Simple, workmanlike and appropriate; giving identity to community and counterpoint to environment in a landscape of horizons.

application cover the gamut of environmental contexts and aesthetic conditions. However, they may be divided conveniently into two broad streams: one is focussed on urban development and is rooted in the design tradition of intuitive expertise, but with important extensions to cover public concerns at the level of city planning; the other emphasises scenic conservation of rural and wild lands and covers the extended practice of resource management to systematically measure these values.[46] Both research streams, although they remain relatively separate in terms of interest, share a common purpose. The focus is on the visual qualities of the landscape and is motivated by a strong concern for their contemporary deterioration. It is this practical imperative, perhaps more than anything else, which unifies the field and brings abstract matters down to tangible questions of importance to society. The purpose of this section is to emphasise and exemplify the inclusive range of contemporary issues which are at sake.

The reaction to blight, like beauty, may vary but the focus of concern is consistent: it is the kind of environment which is emerging in North America in the last half of the twentieth century. Landscape change is an inevitable and necessary result of the continued process of social and economic development. What is disquieting to many is not just the plastic form and amorphous diffusion of manmade elements, but their deliberate replication.[47] The real measure of the "planned deterioration" of North America is in the erosion of the basic character of the three main types of landscape and their intermixture in a new hybrid form.

The aesthetic quality of urban development, the everyday surrounds, incurs most criticism. It is seen by many critics as patently bad and becoming worse. Most North American cities are characterised by sterility at the center and sameness in the suburbs. Few downtown areas, which function as design icons, have not been overwhelmed by the upthrow of monoliths, too often crude structures set in stark surrounds, at once alien and assertive.[48] Not everyone would agree with this assessment, and some notable achievements in the restoration of historic areas and new high-density design have been recorded also.[49] But even the cities which retain a reasonable degree of both visual continuity and human scale are threatened by the continued inroads of glass and concrete.[50] In those older neighbourhoods characterised by real deprivation, aesthetic concerns about slum conditions are a low priority. At the edge of cities, however, the outspread of expensive, low density residential areas frequently assumes a monotonous and non-descript pattern which is superimposed on, not adjusted to, the character of the landscape. The end results of the combined efforts of the "development" industries in many parts of western Canada, for example, are housing subdivisions that are neither attractive nor affordable, being typified by inconsistency within standardization rather than variety within unity.

FIGURE 8,1
New Orleans:
The residential
streets of the
Vieux Carre
speak of an
identity and
cohesion that
is rarely found
in modern
neighbourhoods.

FIGURE 9,1
Learning from
Las Vegas: As the
archetype of the
strip, is it the
Babylon of the
Desert, a symbol of
all that is ugly,
or the linear
equivalent of the
Piazzas of Rome as
Venturi asserts?

Beyond the city, urban sprawl has created a sub-landscape; its very name and existence denotes the absence of clear pattern. The rural-urban fringe is neither town nor country, but a transitional zone of conflicting elements and fragmented ribbon development.[51] It is the North American visual subtopia — the archetype is the ubiquitous highway strip, as much as the congeries of unwanted activities, such as landfill sights, borrow pits and junkyards. As architectural form, the strip is "other-directed", commercially geared to attract the attention of the passing motorist.[52] Businesses compete to intrude into the field of vision and the result is the familiar jumble of garish facades, neon lights and exaggerated signs. This kind of highway fronting now occurs along the approaches to all urban centers, large and small, and extends in other forms a considerable way into the surrounding countryside. It means that the open landscape is increasingly seen through a roadside screen of fringe elements, the aesthetic effects of which are out of all proportion to the amount of land actually occupied.[53]

The rural landscape also is being modified in ways that often reduce its aesthetic qualities. Not all of these are attributable to the urbanisation of the countryside (and nor is the latter always visually detrimental). Modern agricultural economics and technology, based on chemicals as well as machinery, also has an assortment of side effects.[54] A new farmscape, one of less dramatic pattern, seems to be emerging as a result of the abandonment of soil and water conservation practices. Wetlands and vegetation, including ponds and shelter belts established during the drought conditions of the thirties, are being steadily drained and cleared. Rectangular crop fields devoted to monocultures are also replacing contoured terraces, even in lands sensitive to erosion.[55] Here the loss of visual amenity, of course, is symptomatic of serious long term costs of resource productivity.

Similar patterns and processes of landscape change are occurring in the wildlands of North America, although the philosophy and practice of visual and ecological resource management is gaining quite wide currency in the public agencies responsible for administering these areas. This is the case especially in the United States, and perhaps is best developed in the national Forest Service, being exemplified in programs to minimise the aesthetic impact of sustained yield timber harvesting.[56] Mineral resource extraction, relying on huge strip mining machinery, is even more devastating to the visual qualities of wild landscape than clear cutting. Its impact to date has been comparatively localised, but the continued energy crisis opens the threat of massive development of public lands of the western states and provinces.[57] Surface restoration of disturbed sites and other degraded lands is still a relatively experimental practice, particularly in slow growth alpine and sub-alpine regions where scenic impacts are likely to be most serious. An even more widespread threat to the landscape qualities of natural (and rural) areas is the explosion of recreational use.[58]

The problem here is that the scope and scale of demand, especially for increasingly sophisticated support facilities and new opportunities for machine based activity, threatens to erode the scenic qualities which visitors are there to experience by creating the urban-type conditions they are trying to escape.

The reasons for the wholesale decline of aesthetic and other environmental values, in other words, lie deep in our material and technological society. Blight is pervasive, Lewis has noted, because of profit motivation, because of political indifference and because of public apathy[59]—the cautionary tale in the preface of this volume reveals also that it is an insidious condition which is easily overlooked and adapted to. Analysis can help: only when the insights from research become formalised and explicit can aesthetic conditions be taken into account in policy and planning decisions in the same way as are other social and economic values.[60] But this will only take us so far. Ultimately, the only corrective action lies in an adjustment in our contemporary lifestyle. This is the argument developed in Ted Relph's paper; he provides a trenchant critique of the societal roots of visual blandness in built landscapes, and presents some thoughtful recommendations for long term change.

THE SCOPE OF REVIEW

The evolution of environmental aesthetics as a field of study is closely related to the changes that are occurring in the contemporary scene. It had its origins in the early sixties when a number of commentators began to examine the visual landscape. Most of them did not like what was happening and outrage was a common initial response. Anger and polemics have given way to sustained concern and careful study. The scope of review is now multi-dimensional in theme and approach.

Some of the key issues in the theory and practice of environmental aesthetics have been outlined in this introduction and are expanded in the essays which follow. The theoretical and philosophical contributions found in the papers by Appleton and Relph have been noted already. It only remains for us to make the obvious point that their respective conceptualisations on why we appreciate landscape and why it is deteriorating provide important general perspectives on the field. By contrast, the other papers explor representative avenues of approach which are followed in the study of environmental aesthetics. Landscape analysis is concerned with the classification and inventory of the visual characteristics of regions and sites on the basis of formal criteria of physical quality. Burton Litton analyses the theoretical and practical issues associated with its application to wildland areas. A second direction, based on environmental psychology, focuses

17

on aesthetic response and covers the scenic preferences and visual assessments of the lay public. Douglas Porteous provides a searching comparison of this work in relation to professional analyses and reviews the implications for planning and conservation of the urban environment. A final approach concentrates on scholarly examination of literature and art to uncover the aesthetic norms that play an influential role in determining broad cultural patterns in taste for landscape. The pair of papers by Ronald Rees and Richard Harrison respectively examine the way painters and writers portray and evoke the Canadian prairies, a distinctive region where the dimensions of space and scale often overwhelm the conventional sensibilities of even long term residents.

The collection of papers provides an account of progress to date in realising the potentials and dealing with the problems of this complex and fascinating field. It records both how we may organize and develop our sensitivities to the aesthetic quality of the environment and suggests directions for policy and action for protection or improvement. It also points to the barriers which stand in the way of meeting these ideals. The difficulties we encounter in making the concept of the aesthetic operational in relation to environment, as much as anything else, appear to be a consequence of divisions about the importance of man and nature and the respective roles of science versus humanism and expert versus layman in analysis and action. At the end of the volume we discuss further some of these dichotomies.

REFERENCES

1. LYNCH, K., *The Image of the City*. Cambridge, Mass.: The MIT Press, 1960, p. 1.

2. APPLEYARD, D., LYNCH, K. and MEYER, J.R., *The View from the Road*. Cambridge, Mass.: The MIT Press, 1964.

3. THE PRESIDENT'S COUNCIL ON RECREATION AND NATURAL BEAUTY, *From Sea to Shining Sea*. Washington, D.C.: United States Government Printing Office, 1966, pp. 25-27.

4. See NASH, R., *The American Environment: Readings in the History of Conservation*. Reading, Mass.: Addison-Wesley, 1976.

5. McHARG, I., *Design with Nature*. New York: Natural History Press, 1969.

6. An insightful review and discussion of this question is contained in KATES, R.W., "The Pursuit of Beauty in the Environment", *Landscape*, 16, 1967, pp. 21-25.

7. TUNNARD, C. and PUSHKAREV, B., *Man-made America: Chaos or Control?* New Haven: Yale University Press, 1963, p. 3.

8. LEWIS, P.F., "The Geographer as Landscape Critic," in LEWIS, P.F., et. al., *Visual Blight in America*. Washington, D.C.: Association of American Geographers, 1973, pp. 1-8.

9. CARLSON, A.A., "On the Possibility of Quantifying Scenic Beauty", *Landscape Planning*, 4, 1977, pp. 131-172; esp. at pp. 151-166. These skills were cultivated by the great landscape painters of the past. Constable's clouds and Ruskin's mountains, for example, were based on detailed and accurate studies. TUAN, Y.F., "Topophilia, or Sudden Encounter with the Landscape", *Landscape*, 11, 1961, p. 32. It appears to be correspondingly more difficult for physical and human geographers to extend and refine their analytical skills to encompass landscape aesthetics. Two reviews of this problem, which ideally should be read in sequence to illustrate the degree and scope of recent progress, are: DARBY, H.C., "The Problem of Geographical Description", *Transactions, Insti-*

tute of British Geographers, 30, 1962, pp. 1-73; and APPLETON, J.H., Landscape in the Arts and the Sciences. Hull: University of Hull, 1980.

10. See ARTHUR, L.M., DANIEL, T.C. and BOSTER, R.S., "Scenic Assessment: An Overview," Landscape Planning, 4, 1977, pp. 109-130; and LYNCH, K., Managing the Sense of a Region. Cambridge, Mass.: The MIT Press, 1976.

11. A comprehensive discussion of this aspect of the field is in TUNNARD, C., A World with a View: An Inquiry into the Nature of Scenic Values. New Haven: Yale University Press, 1978.

12. See DEARDEN, P., "Landscape Assessment: The Last Decade", Canadian Geographer, 24, pp. 316-325.

13. APPLETON, J., "Landscape Evaluation: The Theoretical Vacuum", Transactions, Institute of British Geographers, 66, 1975, p. 122.

14. This capsule definition of art is from LANGER, S.K., Feeling and Form: A Theory of Art. New York: Charles Scribner's Sons, 1953, p. 40. Comparable definitions can be found in other similar theoretical work on the nature of art and the aesthetic.

15. ZIFF, P., "Reasons in Art Criticism" in KENNICK, W.E., (ed.), Art and Philosophy. New York: St. Martin's Press, 1964, p. 620.

16. The tendency to limit the aesthetic to art has been discussed and criticized in a number of places. See, for example, HEPBURN, R.W., "Aesthetic Appreciation of Nature", in OSBORNE, H., (ed.), Aesthetics in the Modern World. London: Thames and Hudson, 1968, pp. 49-66; or ROSE, M.C., "Nature as an Aesthetic Concept", British Journal of Aesthetics, 16, 1976, pp. 3-12.

17. Both the "vast, wild and unfinished" American landscape and the "small, picturesque and settled" English countryside reflect the repeated interplay between basic physical features and persistent cultural attitudes. LOWENTHAL, D., "The American Scene", Geographical Review, 58, 1968, pp. 61-88; and LOWENTHAL, D. and PRINCE, H.C., "The English Landscape", Geographical Review, 54, 1964, pp. 309-346.

18. For relevant discussions see TUAN, Y.F., "Ambiguity and Ambivalence in Attitudes Toward Environment", *Annals, Association of American Geographers*, 63, 1973, pp. 411-423; BUNKSE, E.V., "Commoner Attitudes Towards Landscape and Nature", *Annals, Association of American Geographers*, 68, 1978, pp. 551-566; and RUBIN, B., "Aesthetic Ideology and Urban Design", *Annals, Association of American Geographers*, 69, 1979, pp. 339-361.

19. HOSPERS, J., *Meaning and Truth in the Arts*. Chapel Hill: University of North Carolina Press, 1946, pp. 12-13.

20. LYNCH, *op. cit.*, (reference 1), p. 119.

21. This includes notably both the design analysis systems involving professional judgement and the various techniques for rating public and expert assessments. Prominent respective examples are LITTON, R.B., *et. al.*, *Water and Landscape*. New York: Water Information Centre, 1974; and CRAIK, K.H. and ZUBE, E.H., eds., *Perceiving Environmental Quality*. New York: Plenum Press, 1976.

22. TUAN, Y.F., *Topophilia*. Englewood Cliffs, N.J.: Prentice-Hall, 1974, p. 94. The author is one of the leading exponents of this approach and the book perhaps is the single richest source of his thinking.

23. This point was made originally by Lowenthal in regard to the broader field of environmental perception: it remains valid with respect to the study of aesthetics. LOWENTHAL, D., "Research in Environmental Perception and Behavior: Perspectives on Current Problems", *Environment and Behavior*, 4, 1972, pp. 335-6.

24. The whole point of the survey exercise is potentially undermined if a meaningful distinction is not made between ratings of quality and preferences for landscape. CRAIK, K.H., "Psychological Factors in Landscape Appraisal", *Environment and Behaviour*, 4, 1972, pp. 256-257. Empirical evidence both supports and contradicts the argument that considered judgement and personal preference are different. Compare, for example, RABINOWITZ, C.B. and COUGHLIN, R.C., *Analysis of Landscape Characteristics Relevant to Preference*. Philadelphia: Regional Science Research Institute, Discussion Paper No. 38, 1970, and DANIEL, T.C. and BOSTER, R.S., *Measuring Landscape Aesthetics: The Scenic Beauty Estimation Method*. Ft. Collins, Colorado: U.S. Forest

Service, Research Paper RM-167, 1976. It is apparent also from these discussions that differences and similarities may be related to the methods employed.

25. *See* CLARKE, K., *Landscape into Art*. London: John Murray, 1971.

26. The literature on this topic is voluminous. For a brief but interesting account of this tradition in relation to venacular architecture, one that is relevant to later arguments, see RAPOPORT, A., *House Form and Culture*. Englewood Cliffs, N.J.: Prentice-Hall, 1969, pp. 1-17.

27. The holistic frameworks of geographical enquiry emphasise the complexity of the interactions between man and environment over space and time. Renewed attention is being paid to the visible and experential aspects of place and landscape as part of a broadening geographical paradigm. *See* BUTZER, K., ed., *Dimensions sions of Human Geography*. Chicago: University of Chicago, Department of Geography, Research Paper 186, 1978.

28. MEINIG, D.W., ed., *The Interpretation of Ordinary Landscapes*. New York: Oxford University Press, 1978.

29. TUNNARD and PUSHKAREV, *op. cit.*, p. 8.

30. For an exemplification see ZARING, J., "The Romantic Face of Wales", *Annals, Association of American Geographers*, 67, 1977, pp. 397-418.

31. GLACKEN, C., *Traces on the Rhodian Shore*. Berkeley: University of California Press, 1967.

32. SADLER, B., "Mountains as Scenery", *Canadian Alpine Journal*, 57, 1976, pp. 51-53.

33. NASH, R., *Wilderness and the American Mind*. New Haven, Conn.: Yale University Press, 1973.

34. O'RIORDAN, T., *Environmentalism*. London: Pion, 1976, pp. 4-7. A strict interpretation of this ideology means that preservation policy for natural areas is restrictively defined and scientifically determined. There is major division of opinion among conservationists

over this interpretation. It is exemplified by the administrative wrangling and political lobbying that is taking place over the United States heritage legislation now before Congress. See FRONDORF, A.F., *et. al.*, "Quality Landscapes: Preserving the National Heritage", *Landscape*, 24, 1980, pp. 17-21.

35. For an interesting discussion of this question, see HODGSON, R.W. and THAYER, R.L., "Implied Human Influence Reduces Landscape Beauty", *Landscape Planning*, 7, 1980, pp. 170-179.

36. TUAN, Y.F., *Man and Nature.* Washington, D.C.: Association of American Geographers, Commission on College Geography Resource Paper No. 10, 1971, pp. 33-37.

37. NAIRN, I., *The American Landscape.* New York: Random House, 1965, p. 3.

38. WARNER, S.B., *The Urban Wilderness.* New York: Harper and Row, 1972, p. 266.

39. STRAUSS, A.L., *Images of the American City.* Glencoe: The Free Press, 1961.

40. FRIED, M., "Grieving for a Lost Home", in DUHL, L.J., (ed.), *The Urban Condition.* New York: Simon & Schuster, 1963, pp. 151-171; FIREY, W.I., "Sentiment and Symbolism as Ecological Variables", *American Sociological Review*, 10, 1945, pp. 140-148.

41. LYNCH, K., *What Time is this Place?* Cambridge, Mass.: The MIT Press, 1974.

42. VENTURI, R., BROWN, D.S., and IZENOUR, S., *Learning from Las Vegas.* Cambridge, Mass.: The MIT Press, 1972. See also BANHAM, R., *Los Angeles: The Architecture of the Four Ecologies.* London: Penguin, 1971.

43. A critical analysis of psychological theory and research on environmental aesthetics is in WOHLWILL, J.F., "Environmental Aesthetics: The Environment as a Source of Affect", in ALTMAN, I. and WOHLWILL, J.F., eds., *Human Behaviour and Environment*, Vol. I. New York: Plenum Press, 1976, pp. 37-86.

44. APPLETON, J., *The Experience of Landscape*. London: John Wiley and Sons, 1975.

45. REDDING, M.J., *Aesthetics in Environmental Planning*. Washington, D.C.: United States Environmental Protection Agency, 1973.

46. LYNCH, *op. cit.*, (reference 10), pp. 5-6.

47. This is discussed, more or less dispassionately in TUNNARD and PUSHKAREV, *op. cit.*, pp. 3-52. For a polemic see BLAKE, P., *God's Own Junkyard*. New York: Holt, Rinehart & Winston, 1964.

48. The criticisms made here of the so called "contemporary cathedrals" of the modern city are not based on size, which after all is a reflection of economic necessity. Rather it revolves, after Burke and others, around their non-style of brutal functionalism. Lacking detail and set in stark concourses, they often contrast harshly with the richness and grandeur of earlier skyscrapers, appearing to have been erected in a design vacuum with little reference to surroundings or people. BURKE, G., *Townscapes*. London: Penguin, 1978, pp. 93-100.

49. One survey of some of the recent accomplishments in urban redevelopment is HALPERN, K., *Downtown U.S.A.: Urban Design in Nine American Cities*. Cincinnati: Watson-Guptill, 1978.

50. See, for example, CIMON, J., "The Siege of the Old Capital by Cement," *Plan Canada*, 18, 1978, pp. 84-86. He is writing of Quebec City, but the comments are equally appropriate to similar urban environments. The basic argument also applies to small towns which are in the process of becoming visually homogenous units of the urban system rather than individual entities rooted in the countryside. See JANKUNIS, F. and SADLER, B., eds., *The Viability and Livability of Small Urban Centres*. Edmonton: The Environment Council, 1979; and LEWIS, P.F., "Small Town in Pennsylvania," *Annals, Association of American Geographers*, 62, 1972, pp. 323-351.

51. For a landscape analysis of the rural urban fringe, see COLEMAN, A., *The Planning Challenge of the Ottawa Area*. Ottawa: Queens Printer, Geographical Paper No. 42, pp. 5-25.

52. JACKSON, J.B., "Other-directed Houses," in ZUBE, E.H., (ed.), *Landscapes: Selected Writings of J.B. Jackson*. Amherst: University of Massachusetts Press, 1970, pp. 55-72.

53. The rural roadscapes of southern Ontario, for example, have recently undergone marked change in visual character. Not only have new urban-architecture forms diffused along their margins, but in the process trees have been removed, road surfaces widened, and utility poles, culverts and the like added. SCOTT, O.R., "Utilizing History to Establish Cultural and Physical Identity in the Landscape", *Landscape Planning*, 6, 1979, pp. 179-203.

54. The processes and patterns of change in traditional farmscapes are analysed in HART, J.F., *The Look of the Land*. Englewood Cliffs, N.J.: Prentice Hall, 1975.

55. These and other examples, such as overgrazed rangelands, are documented in TREFETHEN, J.B., *The American Landscape: 1776-1976, Two Centuries of Change*. Washington, D.C.: The Wildlife Management Institute, 1976, pp. 46-52, 73-76.

56. U.S.D.A. FOREST SERVICE, *National Forest Landscape Management*, 2 vols. Washington, D.C.: United States Government Printing Office, 1973.

57. LUTEN, P.B., "Western Coal Mining", *Landscape*, 24, 1980, pp. 1-2.

58. There is an expanding literature on the environmental impact of wildland recreation. For reviews, see NELSON, J.G. and BUTLER, R.W., "Recreation and the Environment", in MANNERS, L.R. and MIKESELL, M.W., (eds.), *Perspectives on Environment*. Washington, D.C.: Association of American Geographers, 1974, pp. 290-310; and LUCAS, R.C., "Impact of Human Pressure on Parks, Wilderness and Other Recreation Lands", in HAMMOND, K.A., *et. al.*, (eds.), *Sourcebook on the Environment*. Chicago: The University of Chicago Press, 1978, pp. 221-240.

59. LEWIS, *op. cit.*, (reference 8).

60. LYNCH, *op. cit.*, (reference 10).

PLATE 3 Prospect and Refuge: Invitations to Exploration along the Oregon Coast.
B. Sadler Photo ▶

2 PLEASURE AND THE PERCEPTION OF HABITAT: A CONCEPTUAL FRAMEWORK

Jay Appleton
University of Hull

INTRODUCTION

". . . Contemporary writings on aesthetics attend almost exclusively to the arts and very rarely to natural beauty."[1] So wrote R.W. Hepburn some ten years ago, and in a paper published in 1976 Mary Carman Rose made it clear that the situation by then had not radically changed, though in making a plea for the restoration of "nature" as a proper subject for aesthetic study she was in fact helping to change it.[2] "Nature," of course, is not the same as "landscape." However, non-specialists in philosophy who are interested in the aesthetics of landscape, and who have looked in vain for a more vigorous lead from the aestheticians, could hardly fail to welcome Professor Rose's trumpet-call to her fellow-specialists.[3]

At the same time, non-philosophers have a particular opportunity to carry out exploratory sorties into the field of speculation of a kind which a professional philosopher might very reasonably hesitate to undertake. A philosopher is, after all, trained to exercise caution in accepting any proposition, to be on the look out for fallacy, to apply stringent tests to whatever explanation claims to be the truth, and to examine critically the language in which it is couched. His technique is more likely to involve the gradual inching forward from what he conceives to be the sound premises of well-attested knowledge rather than to indulge in flights beyond the frontiers of what is proven.

Yet in any search for understanding there is unquestionably a place for adventurous speculation.[4] It is analogous to the preliminary reconnaissance in a military operation and involves a sketching-in of outlines from the best information available at the time, before the terrain is actually occupied. It is a stage that has to be gone through even if the sketch is subsequently demonstrated to be inaccurate. From the standpoint of positions established by others, whether artists, scientists, or philosophers, let us therefore see whether we can detect some rough outlines of what, on closer inspection, might turn out to be a fruitful line of enquiry.

PLEASURE AND AESTHETIC EXPERIENCE

Since philosophers have not been able to define with certainty what is the nature of beauty, and still argue about the differences between aesthetic and other kinds of experience, let us side-step this for a moment and agree, if we can, that some places, some environments, some landscapes are more capable than others of arousing pleasure in those who experience them. This does not mean a philosophical problem has been solved, but we can get on with the job of having a closer look at the phenomenon of "pleasure" and enquiring further into the relationships between those pleasurable feelings which we experience in the contemplation of landscape and those which we derive from other kinds of activity or involvement.

There are three main kinds of pleasurable experience:

1. doing things (for example, dancing, skating, *performing* music);

2. making things (which is a category of doing things, in which some product is created which survives the act of doing. Examples would include sculpting, embroidering, *composing* music); and

3. perceiving things (for example, looking at paintings, smelling roses, *listening* to music).

Many if not most pleasurable activities involve more than one of these categories and a rigid differentiation would be pointless. Smelling roses, for instance, involves "doing" something and "perceiving" something. Nevertheless, even at this crude level of distinction it is clear that, if we are to fit landscape into this scheme, we should expect the pleasure we take in it to be most closely associated with perceiving. If, however, grounds exist for believing that these areas of experience are all related to each other in some meaningful way for which we can find evidence in disciplines other than aesthetics, even if the precise details of our interpretation of such relationships turn out to be false, we may at least have made some progress in working out a methodology through which we shall *eventually* be able to shed some light on the phenomenon of landscape, which constitutes the greater part of our environment and which has so far proved so baffling or of so little interest to aestheticians.

28

It is now roughly half a century since John Dewey so cogently urged an ethological approach to the interpretation of pleasurable experience, minimizing the distinction between aesthetic and other kinds of experience.[5] He did not overlook the dangers of inferring explanations of human from non-human behaviour, but these dangers are no greater than those of running away from the whole undertaking for fear of making mistakes. If we make erroneous explanations they can subsequently be rectified; if we make no explanations at all we can certainly make no progress. Let us therefore leave aside for the moment the question of how far animal behaviour may be invoked in explanation of human behaviour and consider simply "behaviour" in relation to environment.

BEHAVIOUR AND ENVIRONMENT: A CONCEPTUAL FRAMEWORK

In interpreting the diagram (Figure 1,2), which sets out certain behavioural relationships, it is essential, first, to understand that the activities listed are merely *examples* and do not constitute a comprehensive catalogue, and, secondly, to note that in the present context the system of relationships is proposed hypothetically. The evidence for these relationships is to be sought in the behavioural sciences and cannot be fully discussed here for want of space. The purpose of the diagram is to suggest a conceptual framework within which we can approach the problem of relating the environment, natural or man-made, to other objects of aesthetic experience; it is *not* an assertion of proven facts.

The activities listed in Column 2 are examples of activities which afford some degree of satisfaction to the performer. Some, for example breathing, may not immediately spring to mind as sources of pleasure, but certainly the inability to perform these activities causes *displeasure*. All these activities have five other things in common.

First, they are *not* the exclusive prerogative of *Homo sapiens* but are shared by at least all the more highly developed members of the animal kingdom, including all the primates. Indeed one would need to go much further back along the evolutionary line to find a single species from whose behaviour-characteristics even *one* of these activities was missing.

Secondly, the activities themselves are all related to the biological survival of the individual or the species or both (Column l). The frustration of some, such as breathing, can cause the death of the individual very quickly, in most species within at most a few minutes. Other activities, such as eating, can be dispensed with for longer periods, in some hibernating animals for several months. Yet others, such as mating, are inessential for the survival of the individual but necessary for the long-term survival of the species.

29

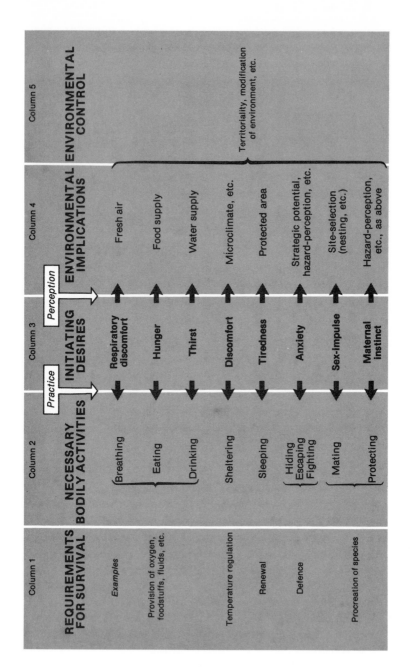

FIGURE 1,2 Environment-Behaviour Relationships.

30

Thirdly, these activities are initiated by desires in the individual (Column 3). Some of these desires can be described by common names such as hunger and thirst. For others we have no simple descriptive words. The operation of the desires is independent of the understanding of any related goal (Column 1) and their satisfaction is not necessarily dependent on the attainment of such a goal. We undertake the associated activities, not because we recognize rationally that they are necessary for survival, but simply because we want to. There is, in short, an association between desire (Column 3) and activity (Column 2) which accounts for a good deal of our behaviour and allows us to recognize, with regard to each activity, certain ranges of performance characteristic of particular species. For instance, cats, horses and different species of birds have "normal" ranges of respiration-rate common to their species: they also have preferences for certain kinds of food, mating habits, methods of expressing aggressive behaviour, etc., all particular to their species.

Fourthly, and following directly from this, the desires (Column 3) together with their associated bodily activities (Column 2) can all be recognized as consisting of two components which we can call respectively "inborn" and "learned." All cats, unlike all rabbits, prefer to assuage their hunger by eating flesh, but some cats grow up with a preference for meat, others for fish. All kinds of fads and fancies may distinguish one individual cat from another, yet it is impossible to avoid the conclusion that there is a common range of preferences which can be observed in all but the most aberrant of cats. These characteristics, physical and behavioural, are inborn; but the behavioural, and in some cases even the physical characteristics may be developed or modified along idiosyncratic lines in each individual as a result of the sum of its experiences. In other words we develop our own preferred methods of gratifying common, inborn desires. We acquire "tastes."

The fifth common attribute of pleasurable activities as exemplified in Column 2 are their environmental implications. Although the idea that we should look for the origins of aesthetic satisfaction in the innate behaviour patterns of animals, including man, as subsequently modified by experience, has received a good deal of serious consideration from philosophers, particularly since the time of John Dewey, they have almost wholly failed to investigate further the *fifth* of the common attributes of pleasurable activities as exemplified in Column 2, namely their environmental implications (Column 4), and it is partly this failure which furnishes the grounds for Professor Rose's complaint.[6] Environmental associations are obviously much stronger in some cases than in others, but collectively they bring the environment within the same explanatory system which enables us to relate pleasurable sensations to bodily activities as part of a comprehensive pattern of behaviour for survival. Some of these activities make very specific demands on the environment. For many animals, for instance, there is a very direct relationship between environmental conditions and the availability of a food supply. The science of ecology has much to say about the principles which govern these relationships. It may seem a far cry from sexual behaviour to environmental optima, but in fact the reproductive process closely involves one of the most particular kinds of link between an animal and its habitat. Many animals will not breed until they have made a nest, and the selection of a nesting-site is a process which can make clearly defined demands on a place in terms of such things as food availability, shelter, concealment, safety from predators, etc. In many species very little variation in practice can be detected between one individual and another in terms of site-selection. A bird's nest will owe its location in *detail* to the experience, acquired preferences and exploratory skill of one bird or pair of birds, but in *general* to the innate preferences displayed by the whole species for a recognizable kind of environmental condition.

Pleasure, Practice and Perception

To proceed to a further examination of the role of "pleasure" in these associations between behaviour and environment, we must look at them in terms of the desires itemized in Column 3 and linked with bodily activities (Column 2) and environmental conditions (Column 4), that is to say in terms of what happens in the brain of the participant. At the cost of some over-simplification we can say that the relationship between Columns 3 and 2 is basically one of *practice*, since the achievement of the satisfaction of the desire (Column 3) depends on the performance of the bodily activity (Column 2). The relationship between Columns 3 and 4, on the other hand, is basically one of *perception*.[7]

In any particular activity practice and perception are inextricably linked. If, for instance, hunger directs that food be eaten, the first step will be to locate food within the environment. Perception will then lead to practice, so that the source of food supply (Column 4) is linked with the performance of the bodily activity of eating (Column 2) thereby satisfying for a time the initiating desire, hunger (Column 3), and incidentally the biological requirement (Column 1).

It may be that an act of environmental perception will initiate the whole process. The sight of some source of danger, for instance, may arouse anxiety (Column 3) which in turn will initiate hiding (Column 2). Further perception will provide the participant with the information on which he can base a decision whether to continue hiding or whether to attempt to escape to a place of greater safety elsewhere, and so on. The alternating and complementary activities of doing and perceiving are equally necessary to the successful accomplishment of the activity (Column 2), the consequent gratification of the desire (Column 3) and the incidental satisfaction of the biological requirement (Column 1), whether or not this is rationally understood. In many activities perception and practice may alternate so rapidly that we think of them as comprising a single experience. For instance, the observation of the road and the manipulation of the steering-wheel of an automobile seem to become fused into the single concept of "driving."

The dangers and difficulties of postulating utility as a criterion of aesthetic experience have played an important part in philosophical discussion on this subject, but in the present context this is a relatively minor problem. If a direct relationship between Columns 3 and 1 had been postulated, we should be in trouble. But this has not been done. Indeed, much stress has been placed on the fact that the activities listed in Column 2 provide the means of satisfying desires (Column 3) *irrespective* of whether some associated ultimate objective (Column 1) is achieved. The neurophysiologist, H.J. Campbell, says:

> The search for food is one of the strongest drives among the lower animals and yet they have not the slightest knowledge that eating does them any good. The ancestral command to activate the pleasure areas[8] ensures that animals explore every means available to stimulate their peripheral receptors and in doing so they inevitably take in nourishment. Eating and drinking behaviours are clear-cut examples of peripheral self-stimulation. Even in man the nutritional aspects are frequently only marginal in terms of conscious motivation. Much of what people eat and drink gives nourishment, but the true marginality of this is revealed when we remember the lengths man goes to in securing these vital materials and the time and money spent in supplying and devouring matter that does not fulfil a nutritional function.[9]

To take an example from a different area of behaviour, failure to achieve procreation does not frustrate the satisfaction of the sex-impulse (Column 3) in mating (Column 2). In a similar way, although the perception of the environment may be related to some ulterior objective, namely the ability to use it advantageously or to assert some measure of control over it, the achievement of *this* objective is not essential to the satisfaction of the desire to perceive.

The attainment of environmental control (Column 5) may involve simply the establishment of some environmental advantage of a strategic kind, as for instance in the phenomenon of "territoriality," in which an animal asserts exclusive rights of occupation over its own "patch," or, at the other end of the scale, the modification, even the transformation of the whole environment. Man is obviously the principal but by no means the only agent of such transformation. One family of beavers, for instance, can create an artificial lake several acres in extent and thereby render more easily attainable, in terms of the methods, habits and practices proper to its species, the various activities (Column 2) which are essential to its survival (Column 1) and to the satisfaction of its several desires (Column 3).

The picture before us, then, is one of a complex system of related impulses, activities and environmental circumstances, in which the conditions necessary for survival are achieved by a sufficient number of individuals for the continuation of the species to be assured. The instigation of the various activities necessary to bring this about is provided by a set of inborn mechanisms, modified by the experience of the individual, which present themselves as "desires" and thereby initiate action. The success of the whole sequence of related operations is frequently dependent on the efficient perception and correct appraisal of the environment, and this itself becomes the object of a "desire" similar to those desires which are listed in Column 3 as initiating the activities listed in Column 2. When environmental perception becomes a purposeful, organized activity, we call it "exploration." Nearly all animals display some desire to explore, and the acquisition of environmental information can very easily prove to be just as much a prerequisite of survival as eating or sheltering or escaping from a predator. We could, in fact, add exploring to the list of activities in the diagram, and again the motivating drive would be simply the pursuit of that pleasure which attends its performance.

Human and Animal Behaviour

It may well be argued that the appraisal of environmental opportunity involves a measure of reason and that this brings us back to the danger of discussing human behaviour in terms of animal behaviour. But the fact is that animals *do* make such appraisals, whether we call it rational or not, and that their survival may depend on it. Aubrey Manning, the ethologist, says:

> A detailed knowledge of the geography of their home
> area will often mean the difference between life and
> death to a small mammal or bird as a predator swoops
> down.[10]

This "knowledge" is acquired by exploration. An animal does not have to be able to explain such knowledge, but it *does* have to be able to act on it.

It is sometimes argued that the differences between human beings and other animals are so fundamental that no valid conclusions can be drawn from comparisons made between them, and it is true that Campbell's theory of behaviour does not deny that these differences are very important:

> . . . the clearest distinction we can make between the subhuman and human is that the human is able to evoke electrical activity in the limbic pleasure areas by processes occurring in the thinking regions of the brain.[11]

But he makes it clear that this capacity is a bonus uniquely available to man and additional to that category of electrical activities in the limbic brain which derive directly from what he calls the "peripheral receptors," (essentially the sense organs of the nervous system) and which are common to human and subhuman alike:

> . . . The basic neural mechanisms of sensory pleasure appear to be the same in man as in the rat.[12]

This is where the argument about the role of cognition and intellectual activity in pleasurable experience comes in. It would be difficult to deny that some pleasure seems to have a cognitive component deriving from a level of intellectual activity which, as Campbell says, is the prerogative of *Homo sapiens*,[13] but it would be even more difficult to sustain the thesis that *all* pleasurable experience is of this kind, and it seems highly probable that at least an important component of the kind of pleasure which comes from the spontaneous perception of our environment must fall within that other category of experience which is ". . . the same in man as in the rat."

Virtually all scientists have now accepted the basic idea that the physical parts of our bodies have evolved in the course of a very long process of adaptation to environmental conditions. It scarcely makes sense to suppose that there has been no parallel transmission of a set of behavioural mechanisms which provide us with the key to using these physical parts. The case that such a transmission does take place and that we are behaviourally as well as physically the heirs of our animal ancestors has been cogently argued by Desmond Morris and others.[14] However, these commentators have so far had more to say about the social than the general environmental context of human behaviour, and it is this last which must principally interest the student of landscape.

It has been argued by some that man has now been civilized for so long that these "animal" traits in his inherited behaviour must surely have died out. But this is not a valid argument. Suppose we allow, say, thirty years per generation as a reasonable average, a hundred generations would take us back to the Bronze Age. Now it is true that much less than a hundred generations of selective breeding would suffice to bring about enormous changes in an animal stock. But we are not here concerned with selective breeding but with natural selection, modified, perhaps, by some controls asserted through social conventions relating to partner-selection, such as are to be found in most societies, but basically free from any semblance of genetic engineering or purposeful breeding programmes. In these circumstances a hundred generations is very little indeed to bring about a complete transformation of the inherited behaviour mechanisms of the species *Homo sapiens*. It is quite irrational to suppose that we are fundamentally different creatures from our primitive ancestors, though of course our behaviour patterns are greatly modified by the very different experiences which we have encountered by virtue of having been brought up in conditions of civilization.

Many of the behavioural characteristics which seem to distinguish man so sharply from other creatures can be seen to arise from the independence which *civilized* man has created for himself. Most of us today can buy our meat from the butcher or fish from the fishmonger; we do not need to catch and kill it ourselves. Neither do we have to be so constantly and keenly aware of all the hazards which confront us as most wild animals need to be, because we have been able to take long-term protective measures against at least some of them, for instance, by setting up police forces, by controlling or eliminating some (but not all) dangerous species of organism, large or small, and by providing efficient devices for temperature regulation such as practical clothing and weatherproof buildings. To a less spectacular degree many domesticated animals are also able to develop behaviour patterns different from those of similar animals in the wild state. In neither case, however, human or non-human, does domestication or civilization bring any fundamental change in the complex system of associations represented in the diagram. This system, or at least the "inborn" component, we continue to transmit genetically from generation to generation, because it is as much a part of our nature as the possession of two eyes and a nose.

Most of the bodily activities listed in Column 2 therefore are still as essential for survival as ever they were, but, as we have become pampered by civilization, some, particularly the defensive activities like fighting, hiding or escaping, are less frequently called on to discharge purposeful roles in ensuring survival, though we might surprise ourselves if we counted up the number of occasions on which we still employ them strategically. The initiating desires (Column 3) do not die out, but they less frequently find opportunities for gratification if they are not so often employed in their original biological function. They therefore demand that other pretexts be found for the continuing operation of these bodily activities, otherwise the pleasure which derives from their performance cannot be achieved. Furthermore, what is really important in the present context is that the environmental assocations (Column 4) continue to operate and we continue to perceive the environment as a theatre for affording or denying the opportunities to hide, to fight, to escape, etc., even though rational man knows very well that, for most of the time, he is no longer in an environmental situation where biological survival depends on either the practice of these bodily activities or the appraisal of the strategic potential of his surroundings to the same extent that it once did. The machinery has been inherited and it insists on being used.

APPLICATIONS OF THE FRAMEWORK

This totality of experience was a central theme in the aesthetics of John Dewey and all the ideas advanced so far are in line with his philosophy. As a criterion of the aesthetic, utility and inutility are irrelevant. What gives us pleasure is the actural performance of those practices which it is in our nature to perform. So the acquisition and practice of bodily skills are desirable objectives in themselves, even if we are not conscious of, or even if we can rationally deny their indispensability for survival.

In amplifying and exemplifying the central role of "experience" in aesthetics Dewey paid more attention to "doing" (activities) and "making" (artifacts) than to "perceiving" (environments), and, as Professors Hepburn and Rose testify, the interests of his successors have continued to show the same bias. But within recent years two important developments have taken place which can afford every encouragement to the view that the time is fully ripe for the balance to be redressed. The first is that the exclusion of nature from aesthetics has now been recognized as illogical in meta-aesthetic terms. As Professor Rose says:

> To be sure, an aesthetic inquiry which is intended to pertain only to artifacts will circumvent aesthetic properties of nature and our responses to them. But it is not a thoroughgoing aesthetic inquiry.[15]

The second is to be found in the vast amount of data made available since Dewey's day by the various sciences which relate behaviour to environment, including ethology, neurophysiology, psychology, geography and that new-fangled hybrid deriving chiefly from the last two, "environmental perception."

Pleasure and Practice

An approach based on the findings of the behavioural and environmental sciences may lead towards a fuller understanding of all kinds of pleasurable behaviour. Thus much recreational activity, for instance, can be seen to afford opportunities for the gratification of desires (Column 3) either by the practice of bodily activities (Column 2) or the perception of the environment (Column 4) or both. Some such activities, for instance running, jumping or swimming, pertain to the perfection of bodily skills (even though the actual form which these skills take may not seem to be directly associated with survival), and clearly fit into Column 2. Their role in terms of survival (Column 1) is to ensure the maintenance of the body in peak condition. Other recreational activities, such as tourism, are more concerned with environmental perception and are equally clearly related to Column 4. Many of these also may be far removed from any immediate function in ensuring survival (Column 1), but, as we have seen from the example of sexual intercourse, the satisfaction of the impulse (Column 3) is by no means dependent on the fulfilment of the related biological need (Column 1).

For some recreational activities, therefore, the environment is relatively unimportant. The boxing-ring and the swimming-pool do not need to be endowed with those properties which we admire in a well-laid-out garden or an alpine valley. Such environmental adjuncts provide an enjoyable bonus but their absence does not destroy the satisfaction to be derived from the central athletic activity.

Other sports directly exploit the pleasure-giving potential of nature by intimately incorporating the perceptual experience of the natural environment within the activity itself. Skiing, sailing, mountaineering and underwater exploration would furnish examples. Indeed all kinds of exploration, from caving to simply going for an afternoon stroll, have an important recreational significance. The new sport of orienteering perhaps pushes the integration of Columns 2 and 4 to the farthest limits so far achieved in any recreational activity, and certainly brings into action also Campbell's uniquely human "two-way traffic between the limbic pleasure areas and the higher regions"[16] of the brain, while those traditional and long-established pastimes of the English country squire—hunting, shooting and fishing—acquire a more understandable significance when related to biological survival within the framework of the diagram. The capacity to perceive an environment critically and selectively, the skill to formulate a plan to wrest the maximum strategic advantage out of it, the physical prowess necessary to put that plan into effect and the final demonstration of superiority in field-craft which is implicit in the capturing or killing of a quarry deeply involve a number of related desires and afford the opportunity for gratifying them both by practice and perception. The predator-turned-sportsman *may* crown his achievement by eating what he has caught, in which case he adds another pleasureable experience and satisfies another desire, but many an English fisherman, believe it or not, is content to throw his catch back into the water, and no traditional English foxhunter would go so far as to eat the fox! The shooter may even substitute a clay pigeon without wholly losing the satisfaction of the experience. Indeed all sorts of modification may be introduced, such as the substitution of photographing for capturing or the concentration on certain parts of the perceiving-killing sequence to the exclusion of others.

It can be seen, then, that Dewey's experiential approach to aesthetics can derive not only a new impetus but also a new direction from the advances which have been made in the behavioural and environmental sciences. Even if the relationships suggested in the diagram must be treated as hypothetical, they do perhaps point the way towards a *modus operandi* by which aestheticians may find meaningful approaches to the problems of bringing back "nature" to an important position within their discipline while the rest of us may be guided towards new ways of interpreting landscape aesthetically.

Pleasure and Environmental Perception

There would not be space here to work out in detail the implication of even the most important aspects of the perceptual relationships between creatures and their environments, natural or man-made, but I have elsewhere suggested a possible approach called "Prospect-Refuge Theory."[17] The part of the diagram which seems to hold out the best promise for the interpretation of environmental aesthetics is that which pertains to "defence," partly because it involves the appraisal of environmental objects as discharging functional roles in a strategic context,[18] but more importantly because the perceiving mechanisms in this group operate with more immediacy than those which relate to any other activities. They have to. A creature which is hungry may have hours if not days in which to discover the means of satisfying that desire. A creature which is threatened by some environmental hazard may have only a split second in which to save its life. It is a well-established fact in psychology that the mechanisms of perception provide for the immediate arousal of a creature's awareness on receipt of a danger signal, to the temporary exclusion of concentration on any other activity.

In the development of aesthetic sensitivity to our environment Prospect-Refuge Theory postulates a particular function for this highly sensitive apparatus for hazard-perception and hazard-avoidance, in which the twin activities of *seeing* and *hiding* play the key roles. By classifying environmental objects and environmental conditions in terms of their capacities to facilitate or impede these funcitons we can set up a typology of landscapes in terms of "prospect" (seeing) and "refuge" (hiding and sheltering). From this beginning we can make some progress towards explaining the satisfaction we derive from our perceived environment, whatever the medium through which it may be encountered (and incidentally towards restoring "nature" to a central position within the enquiry, not because it is "nature" as opposed to "artifact," but because it is the basic raw material of that environment to which our perceiving apparatus remains attuned).

From the vast potential wealth of environmental experience which could serve to illustrate the practical application of the theory, one example must suffice. The activity of *exploring* is, as previously noted, a master-activity by which the creature informs itself of the environmental opportunities for achieving all its other desires. Here again, "pleasure" is the motivating drive; we explore because we *enjoy* it. Environmental conditions which confront us with opportunities to do this, or which symbolically suggest such opportunities, set in motion the desire-achievement-pleasure syndrome. In short, they stimulate our perceiving equipment to do the job it evolved to do by the Darwinian process of natural selection.

This may be illustrated by reference to three constantly recurring morphological arrangements in which the exploratory desire is aroused. The first may be called the "deflected vista."[19] It occurs when a vistal channel or passage (that is to say one bounded by lateral screens or, in the landscape architect's term, "masses") is bent or "deflected," as in a curved path flanked by opaque bushes or as in a curved street, thereby stimulating speculation on what lies beyond the point of deflection. The material of which the bounding walls are composed may be of an infinite variety of substances, natural or artificial, and on very different scales. The challenge to the eye, however, is a continuing common component.

In the second arrangement, a constricted but straight vista in the foreground limits lateral visibility but is succeeded further away by an apparent absence of the bounding lateral screens. An imaginary arrangement of this kind is illustrated in Figure 2,2. The middle picture shows the actual layout and the lower picture the apparent view. There is a strong suggestion that, if the observer were to move forward (from V1 to V2 in the diagram, Figure 2,2), the view would open out into a wide panorama offering further opportunities for indulging the pleasurable acts of seeing, discovering and informing.

The most ubiquitous phenomenon to possess this power of stimulating the expectation of further successful exploratory activity is the horizon, which is by definition the limit of a surface exposed to view from a particular

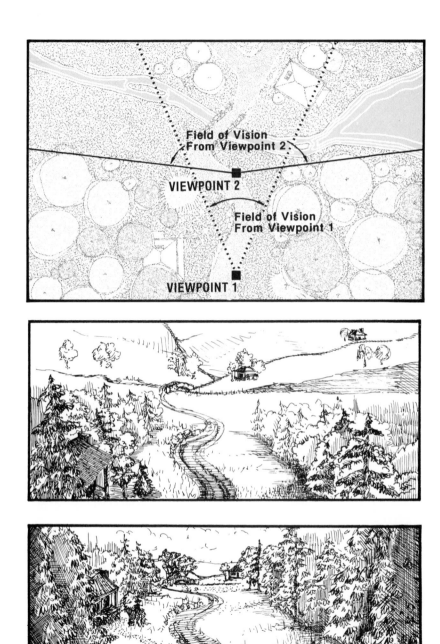

FIGURE 2.2 Imaginary Morphological Arrangement.

41

vantage-point. The evocative power of a horizon may be enhanced in many different ways, for instance by altering the vegetative cover. The horizon can be sharpened by keeping it bare of trees or attention can be concentrated on certain parts of it by leaving bare sections within a generally wooded horizon. It can be made more jagged; objects, natural or man-made, can be raised above it (towers, windmills, ruined castles commonly discharge this function in classical landscape painting). In particular, emphasis can be achieved by placing bright light behind it and even introducing a canopy in the form of a cloud-ceiling to accentuate the "prospect" value of the strip of sky immediately above it.[21]

CONCLUSION

At this point, let us return to Professor Rose's plea for re-introduction of nature into the mainstream of aesthetic inquiry. In the present context we do not need to share too deeply her concern for the proper definition of "nature" because the natural and the artificial are both subsumed under the name of "environment." We have not said, "Let us for a change ignore human activities, like ballet-dancing, and artifacts, like Grecian vases, and instead examine nature." Had we done so we should certainly have needed to be very particular about our definitions of these categories because it would be a principal part of our objective to distinguish between them. But rather we have said, "Let us examine *everything* of which we have experience, including activities, artifacts and environments, natural as well as man-made." We can then see how "pleasure" emerges from our examination and we allow the natural and the artificial to find their own levels within the argument.

Our concern so far has been with the generally accepted fact that people express preferences for different kinds of landscape and with the search for some way of linking "preference" with scientific laws. Allen Carlson, having cited several grounds for scepticism about the validity of the various techniques which have been employed for assessing the "values" of landscapes, concludes his paper with the assertion that, even if the techniques were valid up to the point at which they sought to measure preference-values, they would fall down at that stage at which preference-values were to be translated into *aesthetic* values, because the argument begs a very big question about the relationships between these two sorts of value. For the philosopher this is no doubt the crux of the whole matter, but in the present context it is my intention to stop the discussion short of it, and not to impair the validity of the argument by attempting to achieve too much all at once.[22] Anyone who subscribes to a hedonistic theory of aesthetics may

42

well feel able to go further and accept that this enables us to propose a definition of beauty which is grounded in the laws of the behavioural and environmental sciences along the lines which I have suggested. But I am well aware that many aestheticians would not subscribe to any of the hedonistic theories which have so far been advanced, and for that reason the establishment of a link between the laws of the behavioural sciences and an acceptable definition of "beauty," "aesthetics," "aesthetic values," etc., would present another problem. Furthermore the preoccupation of most modern aestheticians seems to be with "art," and the present discussion has scarcely made more than passing reference to that subject.

As an enquiry into the basis of landscape *preference*, this paper may open up certain lines of thought which will go a little way towards relating observed "taste" to scientific explanation. But as a contribution to *aesthetics* the most that could be claimed for it is that it has attempted to clear the ground somewhat, so that we can see whether there are any indications of paths which might enable us to bring back "nature" into the forefront of aesthetic discussion, where it used to be two hundred years ago but from which it has for so long been banished. That, however, is a task for the philosophers. Naturally I should be delighted if someone better qualified than I were tempted to pursue these issues further.

REFERENCES

1. HEPBURN, R.W., "Aesthetic Appreciation in Nature", in OSBORNE, H., (ed.), *Aesthetics in the Modern World*. London: Thames and Hudson, 1968, p. 49.

2. ROSE, M.C., "Nature as an Aesthetic Object", *British Journal of Aesthetics*, 16, 1976, pp. 3-12.

3. As a recent paper by Allen Carlson demonstrates, when philosophers become interested in what other specialists are doing in the field of environmental preference and the aesthetic implications of landscape, their comments and criticisms are highly pertinent and indeed raise fundamental questions which cannot be ignored. See CARLSON, A., "On the Possibility of Quantifying Scenic Beauty", *Landscape Planning*, 4, 1977, pp. 131-172.

4. See, for instance, APPLETON, J., *The Poetry of Habitat*. Hull: University of Hull, Department of Geography and Landscape Research Group, 1978.

5. DEWEY, J., *Experience and Nature*. New York: Norton, 1929; and *Art as Experience*. New York: Capricorn, 1934.

6. See, for instance, JENKINS, I., *Art and the Human Enterprise*. Cambridge, Mass: Harvard University Press, 1958.

7. Perception is, of course, simply one category of behavioural practice. The word "practice" as used here refers to *other* categories of activity.

8. For an explanation of this and other terms in this passage see CAMPBELL, H.J., *The Pleasure Areas*. London: Eyre Methuen, 1973, which proposes a theory of behaviour in which a central role is accorded to electrical impulses in certain parts of the brain designated as "the pleasure areas".

9. CAMPBELL, *op. cit.*, pp. 131-132.

10. MANNING, A., *An Introduction to Animal Behaviour.*London: Edward Arnold, 1972, p. 190.

11. CAMPBELL, *op. cit.*, p. 71.

12. CAMPBELL, *op. cit.*, p. 110.

13. The fitting together of information and the building up of meaningful images is seen by Peter Smith as a major part of the source of satisfaction to be obtained from the perception of the urban environment. See, for instance, SMITH, P.F., *The Dynamics of Urbanism.* London: Hutchinson, 1974; and *The Syntax of Cities.* London: Hutchinson, 1977.

14. See, for instance, MORRIS, D., *The Naked Ape.* London: Cape, 1967; and *The Human Zoo.* Toronto: Clarke, Irwin, 1969.

15. ROSE, *op. cit.*, p. 8.

16. CAMPBELL, *op. cit.*, p. 72.

17. APPLETON, J., *The Experience of Landscape.* London: John Wiley and Sons, 1975.

18. Compare the environmental psychologists' concept of "defensible space."

19. APPLETON, *op. cit.*, (reference 17), Chapter 4.

20. *Ibid.*

21. This is the phenomenon which I have called the "sky-dado". *Ibid.*

22. CARLSON, *op. cit.*

PLATE 4   Past Landscapes; Present Realities: New Orleans as a memoir of what we have both kept and lost.   *B. Sadler Photo* ▶

3 THE LANDSCAPE OF THE CONSERVER SOCIETY

Ted Relph
Scarborough College, University of Toronto

INTRODUCTION

The man-made landscapes and built environments that have been created in the last few decades have engendered much criticism and little praise. Of course the modern landscape is not all ugly — there are beautifully cultivated flower beds in neighbourhood parks, the dramatic skylines of office towers, the sensuous curves of highways designed to fit the earth's contours — but these are exceptional moments in a world that declares all too obviously our lack of concern for visual relationships and details. Landscape architects and various civic committees make occasional attempts to do something about this by proposing, for example, that trees be planted along commercial strips or instituting regulations to standardise signs. These are commendable efforts, but they can only be cosmetic, for the source of ugliness in modern landscape lies deep within the personality of our technological and consumer society. In order to create landscapes with qualities that can be extolled and which express a wide range of values there will have to be profound changes of outlook and attitudes. One indicator which suggests the possiblity of such changes is the "conserver society" — a concept of a society based on restraint and frugality rather than consumption. However it is not enough to sit back and wait for this to rise; we must first identify the character and depth of the deficiencies of modern landscapes, and work to teach ourselves to be visually sensitive. Only then can substantial changes occur.

FLATSCAPES

In the production of Muzak, the original music, whether by The Beatles or by Beethoven, is rearranged, rerecorded and electronically processed so that all the major variations in tonal range, in noise level and in rhythm are compressed into a narrow band. The familiar result is melodies which are still recognisable but which are somehow all the same because all the extremes and idiosyncrasies have been suppressed. So it is with most of the landscapes of the present age. An architect, Christian Norberg-Schulz, has described them as "flatscapes" because all the highs and lows of meaning have been lost.[1] Of course, there still are differences between places but they do not seem to matter very much. Here we begin analysis by trying to identify some of the types of flatscapes — private, public and ignored.

Most North Americans live in relatively new suburban dwellings that have no inherited meaning or sense of time. These houses have presumably been selected on the basis of promotion and packaging since most of us lack the expertise to judge them by quality, and in any case qualities and locations are all much the same. The houses have been more or less mass produced by a developer or builder in designs meant to appeal to popular taste, and have been arranged in relation to each other according to Federal Building Codes and local zoning by-laws (Figure 1,3). There have been countless condemnations of this sort of environment, and even a respectable academic like David Riesman refers to "suburbanites" as though we are a species apart.[2] But the people who live in these houses have the same sort of concerns and happinesses as anyone else. In the suburbs children are born and grow up, romances are begun, marriages thrive or break up, people die of cancer and of old age. It only takes part of one lifetime for a considerable attachment of care and emotion to be formed with these houses and neighbourhoods, so it is not surprising that most home owners express their concerns and their identity in some way. In terms of landscape this may take the form of garden ornaments arranged on the front lawn, or careful selection of paint colours or plant arrangements — the styles run the gamut from the trim, undecorated lawn to Snow White with her dwarfish entourage (Figure 2,3).

The aim of this personalisation of property is not to demonstrate conformity, as has sometimes been suggested, but to be different within the imposed constraints of the planners and builders, and within the self-imposed and unspoken conventions of the neighbourhood. The process is just the same in those downtown suburbs where young professionals renovate nineteenth century houses by using lots of wrought-iron, plants and varnished wood. But in neither case is it easy to be genuinely different within all the limitations, especially when almost everyone is following the same debased traditional motifs or Better Homes and Gardens fashions, and has to use mass-produced items like quaint coach lamps or plaster bird baths with three sea-horse legs. So while each house and lot is a valued environment, and its decoration and landscaping are an expression of concern for appearance and for commitment to place, the overall result — like Muzak — is a series of limited modulations around limited themes.

In most cultures there have been public places, such as village greens, where individuals could come together for social or political meetings; there have also been public places created for the people by the rulers or governments in order to encourage their involvement as citizens. In present-day

FIGURE 1,3
Trail Ridge in
Scarborough,
Toronto.
The road width,
sidewalk width,
boulevard width,
setbacks, sideyards,
paved driveways
are specified by
various building
codes and by-laws.

FIGURE 2,3
"Some day my
Prince will come".
Snow White waiting
on a townhouse
doorstep in
Brimley Forest,
Scarborough,
Toronto.

landscapes there seem to be three main types of public places. Those identified by public attendance and popularity are often associated with some interesting or relaxing natural feature like the Wisconsin Dells or the beaches of Miami, but the natural attraction may be overwhelmed by added man-made attractions such as Santa's Workshops and Six Gun Cities (Figure 3,3). Another type of modern public place, one established by corporate edict, is the shopping mall. Commercial strips comprising many small retail outlets might reasonably be understood as a social response to consumer demands, but shopping malls create or at least reorient the pattern of demand. Malls I find to be rather insidious since they usually promote themselves as meeting places, equivalents of the old main street, whereas they are in fact sophisticated profit machines with all the decoration on the inside where the money is to be spent and nothing but blank walls on the outside. They are also little totalitarian commercial states, with their own security forces moving silently and swiftly to eject vagrants, students distributing communist propaganda and other anti-profit elements (Figure 4,3).

The third type of public place is the one established by government designation—either to house the government itself or to establish the benevolence of government. In cities these places are usually ultra-modern and progressive, designed according to an aesthetic which pretends automobiles do not exist and keeps them well out of sight. Such buildings and places are often accompanied by various types of propaganda about open plan offices representing access to government and so on, but there is frequently a serious credibility gap between the written message of the "We do it all for you" variety and the visual message of the awe-inspiring, expensive structures which remind each of us of our relative insignificance and political impotence (Figure 5,3).

This classification of public places could be elaborated, but such expansion is not necessary to further the point that public environments now can have precisely that value which designers wish to ascribe to them. Values and meanings do not develop and flourish, but are invented and imposed. If a corporation or government wants to make somewhere significant all they have to do is to hire a good public relations firm who will contrive an appropriate identity. This can be done, for example, by drawing on some obscure event in local history and elevating it into legend. Or alternatively they can create the entire environment as has been done archetypically at Disneyworld, and on a lesser scale in countless rebuilt forts and pioneer villages, or even in the reconstruction of natural Prairie grassland. In Kimberley, in the Kootenay Mountains of British Columbia, the town fathers decided on a Bavarian identity, put in a stadplatz, gave incentives to people to apply the relevant German facades to their property, and so switched it from a declining mining town to a prosperous bavarianized tourist centre. The problem is, of course, that when personalities for places can be fabulated so successfully and so enjoyably, and especially when this

FIGURE 3,3
Another roadside
attraction.
Live Deer and
Giant Lifeless
Indians in the
Adirondacks,
New York State.

FIGURE 4,3
The "private"
parking lot at
Scarborough
Town Centre.

relieves local economic hardship, it becomes impossible to say what is or is not genuine and significant.

Private houses and most of the sorts of public places mentioned above are meant to be seen—they have been ornamented or landscaped or designed for some sort of visual experience. This is not so for many present-day landscapes, which are meant to be ignored, at least visually. There are countless road intersections, sideyards and backyards of factories and apartment buildings, bits of SLOIP (space left over in planning), entire structures for transmitting telecommunications, which are in full view but have no identifiable visual merits (Figure 6,3). Some of these are so bad they could not have been designed to look worse—a remarkable achievement to have got all the poles and wires and newspaper boxes and weeds and crushed coke cans and discarded fast food containers and buildings and spaces into the worst possible visual arrangement without even trying. I am not referring here to those landscapes of trailer homes and junked cars where people really are poor and have neither the money nor the motivation to make things look good. Nor am I thinking of the genuinely ugly environments of mining or the steel industry or other forms of production. It is the landscapes in the middle of affluence and consumption which no one thinks about or cares about—no man's landscapes—to which I object.

Consider modern international airports, for as a creation of the last thirty years they are especially instructive about present attitudes to environments. their visual and other sensory qualities are both ahuman and anti-nature— vast expanses of windswept concrete, ear-shattering noises, parking lots stacked in raw concrete covered with grime, chain-link fences, buildings of no identifiable style scaled to machines not people, ceaseless reconstruction, spiky metal art objects in inaccessible corners, interminable corridors and huge rooms with rows of seats facing nothing in particular. John Ruskin wrote of rail travel in the nineteenth century that "The whole system . . . is addressed to people who, being in a hurry, are for the time being miserable . . . It transmutes a traveller from a man into a parcel."[3] Railway builders invented parcel people, airport designers made the system scientific, and, unless the system breaks down, parcels seem to be unaware of their surroundings, so there is no good reason why airports should be more than processing centres for flows of aircraft, baggage and people. It seems that both passengers and designers have learned to see and hear only on the superficial level of reading signs and hearing flight announcements. Anything ugly or bland or offensive is suffered silently or not noticed at all.

FIGURE 5,3 Nathan Phillips Square in front of the Toronto City Hall: The politicians' view of the citizen. ▶

"Nature is kind," wrote the English poet laureate John Betjeman, "She causes her creatures to adapt themselves to their surroundings; to certain fish in the deepest part of the ocean she gives enormous eyes with which to pierce the watery deep. To the town dweller of today she has given a kind of eye which makes him blind to the blatant ugliness by which he is surrounded."[4] If, by some chance, he should conquer his blindness he can hardly help but be in awe of the fact that he lives in a society that can create, with enormous investments of time, money and expertise, environments which are best ignored. He might also realise that until this selective insensitivity to landscapes is recognised it is impossible even to discuss what is significant in environments without being arbitrary and hypocritical, let alone to create places of value and meaning.

THE SEARCH FOR A STANDARD

In the eighteenth century an opportunity to build was welcomed because a new building would improve the landscape. Now a proposal for a development is almost always greeted by the reaction—"What a pity they are going to build there." In the eighteenth century, and in most previous ages, there seems to have been a collective vision of the goals of the society and how to achieve them, so that everyone knew what were the qualities of beauty and sublimity. In our age there seems to be no such collective focus. An anarchic relativism is at work in aesthetic and environmental appreciation; modernist architects create elemental forms from steel, glass and concrete, while architectural conservationists believe that nothing of value has been built since 1900; many people can think of nowhere better to live than in a brand-new Colonial style house, but their neighbours opt for a raised ranch house. Modern artists experiment with abstractions that are incomprehensible rubbish to most people, who prefer to watch the realism of Kojak on their colour T.V.s with Mediterranean consoles in the corner of open plan living rooms with Renoir reproductions and day-glo sunsets on felt hanging on their walls.

It is not surprising, then, that there is a quietly desperate search for some consistent standard by which we can assess environments and what we are doing to them. The appeal to authority is pointless since there are no commonly accepted authorities, and anyway the supposed experts will probably disagree, just as they do over the environmental effects of nuclear-power stations. The measurement of public opinion is a popular technique, but of course most public opinion is uninformed on technical issues, and in any case the aggregation of perceptions and opinions will lead directly back to the compressed averages of the "flatscape." Objective and supposedly scientific measures of environmenteal quality would provide a mar-

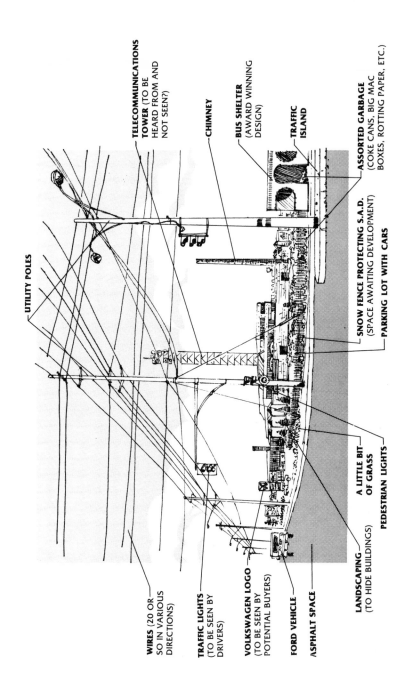

WIRES (20 OR SO IN VARIOUS DIRECTIONS)

TRAFFIC LIGHTS (TO BE SEEN BY DRIVERS)

VOLKSWAGEN LOGO (TO BE SEEN BY POTENTIAL BUYERS)

FORD VEHICLE

ASPHALT SPACE

LANDSCAPING (TO HIDE BUILDINGS)

A LITTLE BIT OF GRASS

PEDESTRIAN LIGHTS

UTILITY POLES

TELECOMMUNICATIONS TOWER (TO BE HEARD FROM AND NOT SEEN?)

CHIMNEY

BUS SHELTER (AWARD WINNING DESIGN)

TRAFFIC ISLAND

ASSORTED GARBAGE (COKE CANS, BIG MAC BOXES, ROTTING PAPER, ETC.)

SNOW FENCE PROTECTING S.A.D. (SPACE AWAITING DEVELOPMENT)

PARKING LOT WITH CARS

FIGURE 6.3 A modern no-man's landscape.

55

vellous way out. Unfortunately, these have about the same relationship to landscape value and meaning as the measurement of alcohol content does to the quality of wines. They are nevertheless easy and undemanding and capable of being applied by anyone regardless of their sensitivity to environments: "For one person who can recognise the loveliness of a look or the purity of a colour," John Ruskin observed, "there are a hundred who can calculate the length of a bone."[5] Attempts to measure environmental and landscape qualities and values will get us nowhere except into greater triviality and pseudo-scientific confusions.

This is depressing and negative stuff, but it will have to get worse before we begin to discern some ways out of our present situation. Two arguments made here can be formulated differently. The remarkable achievements of the eighteenth century architects and landscape gardeners can, in part, be attributed to the visual sensitivity which they taught themselves. "Our sight is the most perfect and most delightful of our Senses" Joseph Addison had written in the *Spectator* of 1712, and he proceeded to show how cultivated seeing could open up all the realms of imagination.[6] This sort of imaginative seeing of landscape was undertaken according to principles of composition, proportion and order which had to be learned, but which were applied with great flexibility and subtlety. In contrast, our century is characterised by what might be called "visual illiteracy." This has been recognised and stated repeatedly. Clough Williams-Ellis said of England in the 1930s: "We are a country of the blind. We do not see and we do not care."[7] More recently, George Nelson, an American instrumental in developing the modern pedestrian mall, has claimed that 90 per cent of North American adults cannot see except in the most primitive sense of recognising their neighbour's dog or a traffic light.[8]

There must be many reasons for this sort of visual incompetence. Ruskin thought speed and mobility had something to do with it: "There was always more in the world than men could see, moved they ever so slowly. They will see no more for going fast."[9] Bookish education must take some blame too; C.S. Lewis observed of the hero in one of his science fiction novels ". . . his education had had the effect of making the things he read and wrote more real to him than things he saw."[10] Television reinforces visual insensitivity to environments not only because it is rarely more than a way of illustrating simple stories with moving pictures, but also because the small screen is particularly poor for showing landscapes or townscapes. But perhaps the most important reason for our visual incompetence is a deep-seated consumer attitude that results from the attempt to found society not on religious or humanist or political grounds, but on the sole fact of economic production. This consumer attitude inclines us to treat everything in terms of its exchange value rather than in terms of its intrinsic worth. So environments are ranked and compared—the Rockies are better than the Appalachians, San Francisco is nicer than Boston—or they are simply fenced off and priced. I have heard of a widely advertised panorama of the Grand Tetons in Wyoming that is entirely hidden by a ten foot high fence. For a couple of dollars you can go through the turnstile and gift store to see the view.

Just after the Second World War, T.S. Eliot made some observations on the state of contemporary culture—and he used the term not in an elitist sense but as an anthropologist might to refer to the entire way of life of a people. His conclusions were not cheerful: "We can," he wrote, "anticipate a period of some duration of which it is possible to say that it will have no culture."[11] By this he meant, I think, that there will be nothing spontaneous or unselfconscious, nothing spiritual, nothing that comes from the grass roots of society. Instead, values and tastes will be given to us by designers and social scientists and advertisers and government officials. I believe I have demonstrated that the evidence of landscape shows that, even if such a cultureless society is not already here, it is dangerously close. Eliot foresaw the end of the cultureless period occurring when values once again developed from the grass roots, from the genuine and deep concerns of people. We cannot solve our problems by adopting the standards of the eighteenth century or of any other age, however much we might admire them, for ours is a mobile and relatively classless society; nor is it enough to plant petunias around the garbage dump and to landscape the nuclear power plant, for, as George Nelson observes, that is like going to Elizabeth Arden to get a cure for cancer.[12] What we have to do is to go back to the beginning, and to rediscover some roots and some fundamental values, and to attempt to match these with our present situation. The alternatives to our visually insensitive non-culture are necessarily radical.

THE EDUCATION OF VISION

Writing on the disfigurement of the English landscape in the 1930s D.H. Lawrence commented that "The point is not whether we can do anything about it or not, all in a hurry. The point is to become acutely conscious of what is happening and what has happened. And as soon as we are really awake to this we can begin to arrange things differently."[13] A first step in becoming "acutely conscious" must be learning to see well. In order to make valuable judgments about landscapes and environments we must observe them carefully and without being arbitrarily selective, and we must think about what we are seeing. This may seem so obvious that it is immediately dismissed as trivial, but seeing is not just a physiological act, it is a faculty that involves emotion and thought, that can be trained or can

atrophy. There are levels of seeing that range from a passive and unreflective looking at objects in order to assess their commercial value or what they
will do for us, through careful and unprejudiced observation and aesthetic
contemplation, to a sensitive apprehension of the identity and meaning of
particular places and environments.

The task that we have to set for ourselves is to strive for the highest
possible sensitivity of vision. For some people this is probably easy and
natural, for others enormously difficult. A good analogy might be learning to play a musical instrument; some individuals can learn to play well,
with considerable depths of interpretation, while others cannot get beyond
some basic pieces no matter how hard they try. But this latter group will
have done enough to appreciate the abilities of those who play better than
they can.

A point to begin the education of vision is with the questions which
Goethe posed for himself as he set out on a journey to Italy in 1786.
"Can I learn to look at things with clear, fresh eyes?" he wrote in his
diary, "How much can I take in at a single glance? Can the grooves of
old mental habits be effaced? This is what I am trying to discover."[14] The
approach he adopted to address these questions was clear and unbiased
observation, regarding everything with impartiality yet without detachment,
so that he could comment equally upon Roman and Renaissance art and
architecture, the faces and dress and customs of contemporary Italians, and
the garbage in the streets of Verona. Judgements and evaluations came
after such observations. Later in his diary he comments on the success of
his method:

> I am in a state of clarity and calm such as I have not
> known for a long time. My habit of looking at and ac-
> cepting things without pretension is standing me in
> good stead and makes me secretly very happy. Each
> day brings me some new remarkable object, some new
> great picture, and a whole city (Rome) which the
> imagination will never encompass, however long one
> thinks and dreams.[15]

This is a strong testimonial for a way of perceiving the world, but we
should be careful to avoid stressing too strongly the merits of unpretentious
seeing. This is only a first step and seeing well necessarily involves some
expenditure of effort beyond this. John Dewey expressed this succinctly—
"to perceive a beholder must create his own experience."[16] Exactly what
this creativity involves and where it will lead I cannot say. I do know that
the rewards of creative vision are not material or pecuniary, but rather have
to do with the quality of one's life. I also believe that without sensitive and

creative vision we will never be able to grasp what is of genuine value to us in the environments in which we live.

THE LANDSCAPE OF THE CONSERVER SOCIETY

Seeing well may enhance our lives as individuals, and it will surely make us more aware of the appalling way in which we treat our landscapes. However, the full realisation of sensitive seeing in making and managing environments requires a social context. Such a context would be provided by what is coming to be known as "The Conserver Society."

Arguments for the stable state economy and ecological dependence of renewable resources that typify the conserver society have been made by economists, ecologists and, a little surprisingly, by the Science Council of Canada. The Science Council report argues that the growing awareness of the fragility of ecosystems, the concern for action on environmental matters and demand for participation in decision-making, the trend toward decentralisation, the realisation that unlimited growth is impossible and the growing sense of constraints to our current economic life-style, all point towards the inevitability of a society which conserves rather than consumes.[17] This is a significant conclusion for even to talk seriously of conservation suggests that something is radically amiss with the principle of consumption on which our society is now based (Figure 7,3). At the same time, some caution is warranted in adopting the official arguments for a conserver society, for while they have attractive overtones of alternative technologies and whole-earth lifestyles they also have undertones of persuading us to accept uncomplaining the coming resource and material shortages and energy cutbacks.

I do not dispute these arguments of the Science Council, they all seem to be very reasonable and sane and necessary, but I also believe that economics and ecology are not a sufficient foundation for a new society. It is the associated changes in social values that are really fundamental. E.F. Schumacher, one of the prime advocates of the economics of smallness and the ecology of restraint, certainly recognised this. He claimed that: "the economic problem is not soluble except on the basis of what I call a metaphysical reconstruction, or if you prefer those terms, a religious or philosophical reconstruction. If we do not sort out our fundamental beliefs we are just drifting."[18]

FIGURE 7,3 A sign of the end of the times: Production Drive is in Progress (!) Park (an industrial subdivision) in Scarborough, Toronto. ▶

Schumacher's second book, *A Guide for the Perplexed*, gives an indication of what is involved in such a metaphysical reconstruction.[19] We should, he maintains, recognise something that has been recognised by virtually every age and culture except ours, namely that there is a heirarchy of being and that some concerns are intrinsically more significant than others. However, we can only know what we have the capacity to know, and if we rely on "the single, colour-blind eye" that is all that has been required to achieve most of the results of modern science, then we can never expect more than explanations of the material world. To develop, or perhaps regain, our understanding of the higher, spiritual levels of human existence we need to exercise *all* our senses as well as we can.

The essential feature of Schumacher's argument is that unless we have a hierarchy of values by which to judge environments and issues and actions we can never do much more than muddle through. An ecologically balanced society with a stable state economy might last longer, but it is scarcely an improvement over our commodity rich and environment poor world if it cannot offer some sense of purpose and fulfilment to the individuals that comprise it. Though this hierarchy is not entirely explicit in his writing, I believe Schumacher has placed at the lowest level of significance the concerns and achievements of the sciences of matter, including classical economics. At a higher level are the concerns of the sciences of life, including botany and biology and ecology. Above these are the values of individual and social identity—in which context it should be noted that J.B. Jackson has suggested that for individuals those landscapes which have been the source of self-awareness are beautiful.[20] The highest values are those which are spiritual, and give a sense of meaning to life, values which the Romantic poets, Renaissance architects and countless others have experienced through natural or man-made environments.

A culture which has such a hierarchy of values will reflect them in its landscapes; the capacity to express what matters most to people will exist and some places and environments will be known to be invested with great meaning. Of course there would not be any ultimate resolution of the problems of which particular values are expressed and by what means they are conveyed—each generation has to do that for itself—but the bases for resolution would be there. And unlike the horizontal and superficial qualities of our flatscapes, there will, in the landscape of the Conserver Society, be a clear vertical dimension to geographical experience.

PRACTICAL IMPLICATIONS

All this no doubt seems very elitist and philosophical. In so far as it is dealing with fundamental issues it has to be philosophical, and if you do try to sidestep philosophy you are almost bound to end up planting petunias around gas stations and garbage dumps, or trying to find superior ways to measure the tonal variations in black and white photos to assess landscape aesthetics. J.B. Jackson once commented that the primary task of environmental design has nothing to do with aesthetics or computers, but is to find ways to design environments where it will be possible to live as free and responsible citizens. These environments may not be beautiful but they would suggest order and justice and in time we would come to see beauty in those expressions.[21] Still we have to begin doing something somewhere and a few suggestions are offered below.

The first and continuing task, as already argued, must be to make ourselves as aware as we can of the qualities of the environments in which we live by teaching ourselves to see them well. But this may be a long and gradual undertaking so we need some more immediate way to assess environments and proposed developments. A possible basis for this is the broad questions Schumacher levelled at any technological development, whether a suburban subdivision or an oil pipeline.[22]

1. How is it relevant to our energy problem? Does it deplete resources, or employ renewable sources of energy?

2. How is it relevant to our ecological problem? Does it at any stage involve adding further pollutants to the biosphere or interfere destructively with ecological processes?

3. How is it relevant to the human problem? Is it a stunt solution for the wealthy or is it addressed to issues of poverty and inequality?

4. How is it relevant to the mental and spiritual health of mankind? Is it mind-killing or enlivening?

Asking these questions seriously, and seeking solutions that would provide positive answers to them, will soon lead to the recognition that large scale developments, conceived and implemented by experts, are both anti-nature and anti-human. We need instead to think on a small scale, to use established solutions rather than invent new ones, to change environments incrementally and gently. The people who live in places and who care about particular environments must be allowed to be involved in their design and maintenance; only through effective involvement will we come to care deeply about environments, and it is only through the investment of human concern that environments will acquire significance. And if all this seems abstract and idealistic remember that these things were done as a matter of course in most parts of the world until only a century or so ago. Furthermore all the bases and many of the details of a self-conscious, small-scale, locally implemented approach to environmental design have been formulated, and applied sucessfully in limited situations, by Gandhi and his followers, by numerous intermediate technologists, by Christopher Alexander and his colleagues at Berkeley, and so on.[23] The point is that we know what to do and how to do it—it is the will that is lacking.

And to the charge that this argument is elitist, advocating a rarified intellectual approach open to but a few, note that the problems of our age and culture have all the characteristics of pesticide pollution—they are invisible, cumulative, deep and complex, and have to be approached on many levels and in many ways. At least one of these levels must be intellectual since so many of our difficulties derive from the scientific world view which pervades our culture. Furthermore I am not proposing some Utopian dream-world in which only the highest and best is found, only that we ought to aim for possibilities other than those represented by the mediocre flatscapes we know and make so well. That there is a heirarchy of values by which we should judge matters will be obvious to anyone who makes the effort to think about it. That we need to do something about the state of our landscapes should be apparent to anyone who chooses to look at them. But really all that is being argued is a very simple idea that surely applies to people in all walks of life, and which was stated perfectly by the architect Sinclair Gauldie: "To live in an environment which has to be endured or ignored rather than enjoyed is to be diminished as a human being."[24]

REFERENCES

1. NORBERG-SCHULZ, C., "Meaning in Architecture", in JENCKS, C., (ed.), *Meaning in Architecture*. London: The Cresset Press, 1969, p. 228.

2. RIESMAN, D., *The Lonely Crowd*. New Haven: Yale University Press, 1950.

3. RUSKIN, J. in HERBERT, R.L., (ed.), *The Art Criticism of John Ruskin*. Garden City, New York: Doubleday, 1964, p. 146.

4. BETJAMEN, J., *Ghastly Good Taste*. New York: St. Martin's Press, 1971, p. 14.

5. RUSKIN, *op. cit.*, p. 4.

6. ADDISON, J., *The Spectator*, 411, 1712, n.p.

7. WILLIAMS-ELLIS, C., *England and the Octopus*. Portmeirion: Penrhyndendraeth, 1928, p. 104.

8. NELSON, G., *How to See*. Boston: Little, Brown and Co., 1977, pp. 2-3.

9. RUSKIN, J., *Modern Painters*. Vol. III. London: Routledge and Sons, 1904, p. 331.

10. LEWIS, C.S., *That Hideous Strength*. London: Pan Books, 1955, p. 56.

11. ELIOT, T.S., *Notes Towards the Definition of Culture*. London: Faber and Faber, 1948, p. 19.

12. NELSON, *op. cit.*, p. 2.

13. LAWRENCE, D.H., *Selected Literary Criticism*. London: Heinemann, 1956, pp. 142-143.

14. GOETHE, J.W., *Italian Journey*. [Translated by W.H. Auden and E. Mayer]. Harmondsworth: Penguin, 1970, p. 38.

15. *Ibid.*, p. 136.

16. DEWEY, J., *Art as Experience*. New York: Capricorn, 1958, p. 54.

17. Science Council of Canada, *Canada as a Conserver Society*. Report No. 27. Ottawa: Supply and Services Canada, 1977.

18. SCHUMACHER, E.F., Interview for Ideas, CBC Radio. October 1977.

19. SCHUMACHER, E.F., *A Guide for the Perplexed*. Toronto: Fitzhenry and Whiteside, 1977.

20. JACKSON, J.B., "The Historic American Landscape", in ZUBE, E.H., (ed.), *Landscape Assessment*. Stroudsburg, Pa.: Dowden, Hutchinson and Ross, 1975, p. 9.

21. JACKSON, J.B., "The Public Landscape", in ZUBE, E.H., (ed.), *Landscapes: Selected Writings of J.B. Jackson*. Amherst: University of Massachusetts Press, 1970, p. 160.

22. SCHUMACHER, *op. cit.*, (reference 18).

23. ALEXANDER, C. et. al., *The Oregon Experiment*. Berkeley: University of California Press, 1975.

24. GAULDIE, S., *Architecture*. London: Oxford University Press, 1969, p. 182.

PLATE 5 The City as Experience: Walking the Freedom Trail through Old (and new) Boston. *B. Sadler Photo* ▶

4 URBAN ENVIRONMENTAL AESTHETICS

J. Douglas Porteous
University of Victoria

In the play of forces that govern the world,
aesthetic ideas rarely have a major role.

Yi-Fu Tuan

INTRODUCTION

In his theoretical, and much-criticized, six-level hierarchy of human needs, Maslow placed "cognitive-aesthetic" needs at the lowest level of importance.[1] The sales of popular non-fiction in the Western world suggest that, although basic physiological needs may now be generally satisfied, most citizens are still much more concerned about affiliation, esteem and self-actualization needs than about aesthetics. And, in the latter area, the personal aesthetics of body and home are more highly regarded than the aesthetics of the public environment. If, as Pawley suggests, we face a largely "private future," the entrenchment of private affluence and public squalor seems inevitable.[2]

Further, in terms of current environmental problems, aesthetic issues are of very low priority, and are frequently considered a mere luxury. This is especially so in North America, where the overwhelming concern has been to obey the Jeffersonian imperative to embrace first the "practical arts," and only later the "decorative" ones. In Lynch's words, "Esthetics is often considered a kind of froth, difficult to analyze, easy to blow away."[3]

On the other hand, the century-old concern for wild landscapes has recently been joined by a growing concern for tangible reminders of the human past. Fake nostalgia booms apart, this movement has seen some admirable initial work in the areas of the conservation, preservation, and rehabilitation of valued landscapes.[4] "Listed" historic buildings in Britain and the "heritage" movement in North America attest to a growing interest in the aesthetic quality of environment, at least on the part of a vocal elite. Moreover, a small but growing body of literature is suggesting that the sensory quality of the environment may have significant effects upon human performance and feelings of well-being. As yet, however, "sensory planning" is in an embryonic state.

In terms of the rural-urban contrast, far greater attention has been paid, by researchers and legislators alike, to the non-urban environment. A formidable battery of techniques exists for the appraisal of the aesthetic qualities of landscapes.[5] It is not my purpose here to question the validity of quantitative assessments of landscapes which, in a most unaesthetic jargon, have become known as "scenic resource values" and "regional recreation facilities." Rather, the overwhelming emphasis on rural and wilderness settings is questioned. It is asserted that this research thrust is essential on account of the rapidly diminishing resources of wilderness and unspoilt countryside. Such resources appear to be eroding at a fast rate because of the impulse of citizens to escape, as often as possible, from their urban dwelling-places to the renewing, re-creative outdoors.

This emphasis on nonurban landscape aesthetics incorporates the common anti-urban bias so tellingly recounted in *The Intellectual Versus the City*.[6] The siting of an open-pit mine or the near-extinction of a small furry mammal is deemed more worthy of concern than the wholesale demolition of obsolete machinery, eighteenth-century warehouses, or Victorian workingmen's row-housing. Although these items may be the concern of urban historians, architectural historians, and industrial archaeologists, the focus of much landscape aesthetics has resolutely been turned towards the nonurban scene.

Perhaps it is forgotten that the Western industrial world is largely an urban world. The mass urbanization of the world's population proceeds apace.[7] A mere 5 percent of the world's population occupied cities of over 100,000 inhabitants in 1900. By 1970 the proportion had risen to 23.7 percent. The estimate for the year 2000 is about 40 percent. Within a generation more than half the global population may be living in major urban agglomerations. Further, in Western societies, urbanization becomes almost total, as the small nonurban population is increasingly subordinated to the economic, social and political styles of the city.[8]

Most urban dwellers have little time to spend outside the urban setting; weekend trips rarely reach beyond the urban economic region. A more balanced landscape aesthetics research posture would devote at least equal time to the assessment of the landscape values of urban areas, with a view to both their understanding and their improvement. Scarce research resources should be redirected from the evaluation of scenic areas which the citizen may see rarely, if at all, during his lifetime. If, instead of ignoring the urban scene, we could improve the citizen's aesthetic satisfaction with it, there might result a lowering of pressure on non-urban areas through a decline in out-of-city discretionary travel. And if recent research findings on the apparent anxiety-producing effects of urban life are validated, the case for a reorientation of research effort towards the everyday urban environment is even greater.

Sceptics may still consider that *Homo economicus* is alive and well and living in suburbia, and exercises his aesthetic capacity only when leaving the city for the countryside. But even the mundane chores of the journey to work and the shopping trip appear to have an aesthetic component. The aesthetic nature of both the shopping area and the route taken to reach it are important factors in consumer decision-making.[9]

In the sections which follow, a nonurban-based theory of environmental aesthetics is compared with an urban-oriented concept. A brief survey of major contemporary approaches to urban environmental aesthetics prefaces a discussion of the importance of the linkages being forged between these approaches. A final section seeks to expand the general scope and role of environmental aesthetic research.

THEORETICAL UNDERPINNINGS

Two of the most recent attempts to generate an explanatory theory of environmental aesthetics are Appleton's *The Experience of Landscape* (1975) and Smith's *The Syntax of Cities* (1977).[10] The former is very much concerned with natural landscapes, but makes explicit reference to applicability within the urban context. The latter is devoted wholly to the urban experience.

Appleton's work emerges from the welter of speculation, during the 1960's, on the biological bases of human behaviour. His essential argument is that our aesthetic reactions to landscape are in part inborn, and "if he is to experience landscape aesthetically, an observer must seek to recreate something of that primitive relationship which links a creature to its habitat."[11] Appleton's habitat theory suggests that both animals and premodern man appreciate landscape largely in terms of survival functions. A further development, known as prospect and refuge theory, is based chiefly on the principle of hunting.[12] Hunters seek vantage points, or prospects, from which to view the prey. The hunted, in contrast, seek hiding-places, or refuges. In essence:

> Habitat theory postulates that aesthetic pleasure in landscape derives from the observer experiencing an environment favourable to the satisfaction of his biological needs. Prospect-refuge theory postulates that, because the ability to see without being seen is an intermediate step in the satisfaction of many of those needs, the capacity of an environment to ensure the achievement of this becomes a more immediate source of aesthetic satisfaction.[13]

For humans, at least, the gap between the biological survival requirements presented by a landscape and the feelings of pleasure derived from the scene is bridged by the symbolic value of environmental props.

A wealth of literary, artistic, and ethological research material is used to support this somewhat zoomorphist argument. Whether the theories are amenable to testing will depend on the ingenuity of future research workers. The evidence provided by Appleton is in no way conclusive; indeed, he wishes "merely to suggest."[14] But his work is almost wholly based upon nonurban landscapes where contour, water, and vegetation, or the lack of these, provide ample scope for the discovery of prospects and refuges. The application of the theories to townscapes is only weakly developed.[15] Buildings are seen, in prospect-refuge terms, as symbolic substitutes for those natural environmental features which allow one to see without being seen. Buildings are modern man's normal refuges; streets, alleys, alcoves and the like provide both prospect and refuge. Unplanned, spontaneous urban growth is likely to be more aesthetically pleasing than the uniformity of planned settlements, whose major lineaments can be comprehended at a glance.

Appleton suggests the lack of an alternative, universally-recognized basis for the aesthetic interpretation of the built environment.[16] It is unlikely that such a model will ever appear, although attempts have been made in this direction. One of the most recent is by Smith.[17] Here the approach is that of neuropsychology. Following recent split-brain theorists, Smith establishes an initial trichotomization of the fore- and mid-brain. The limbic system, seat of the emotions, is complemented by the neocortical left hemisphere, rational, verbal, mathematical, logical, analytical, and deductive, and the right hemisphere, which is holistic, intuitive, spatial, and pattern-recognizing.

Using a wealth of examples, Smith explains how a variety of common urban physical elements creates a complex aesthetic potential. Psychological rewards may be obtained from the townscape via the three neurological processors noted above. The potential for aesthetic satisfaction, however, has increasingly been eroded during the twentieth century. Modernist architecture is "Left-cerebral dominant," minimalist, and thus productive of a myriad indentitowns each graced by a central growth of faceless glass cubes. Smith suggests that contemporary Modernism denies the limbic brain and that we mortify the flesh by starving our right hemispheres.

To satisfy these physiological cravings, tourists flock to the creative jumble of medieval Italian hilltowns or the limbic sensationalism of nocturnal Las Vegas. To Smith, "Las Vegas represents a concentrated eruption of limbic desires, unrestrained by the finer sensibilities of the higher brain."[18] With his cry of "viva vulgarity!" Smith would certainly approve of the recent post-Modernist trend towards more complex, colourful, and allusive architectural forms.

One should be grateful for courageous theoreticians who are willing to take the broad view. Unfortunately, however, both approaches have similar drawbacks. Both draw strongly on ethological and other non-human-based research, yet the problem of extrapolation to man from studies of rats and cats is far from solved. The search for aesthetic universals, or any universals for that matter, is an extremely hazardous endeavour. It is of little value, for example, to speak of universal reactions to angles and corners when some cultural groups, living in a rounded non-carpentered world, fail to perceive these features. Reductionism is alive and well in these ethological and neuropsychological theories. One balks, for example, at Smith's explanation of Brennan's "law" of shopping behaviour as the effect of "a deep-seated psychological pull generated by experience and memory, mixed with deep-rooted tendencies which link us with our urban ancestors in the Mesopotamian and Indus valleys."[19]

Conceptualization of the origins of environmental aesthetics are clearly in the formative stage. In contrast, a great deal of empirical work in the area has already emerged. This work is considered in terms of four general approaches, each distinguished by its particular goals, methodology, and social context.

APPROACHES TO URBAN ENVIRONMENTAL AESTHETICS

Two major philosophical trends have emerged in the social sciences during the last generation. One is the quest for rigour, whereby many social scientists have attempted to become as indistinguishable as possible from their physical scientist colleagues. Quantitative, behaviouralist approaches are indicated. The second is the quest for relevance, pursued with varying vigour by those concerned with applicability, policy orientations, and more extreme radical, Marxist, or humanist views. Rigour regardless of relevance was once pursued with vigour. The more recent trend is towards "relevance with as much rigour as possible."[20]

Using the two criteria of relevance and rigour, four approaches to the study of urban environmental aesthetics have been identified. Relevance, referring to the immediacy of the approach to current environmental problems, is clearly of importance in view of the widespread belief that landscape quality is undergoing considerable decline through the multi-dimensional assaults of urban-industrial civilization. Rigour, which refers to scientific theory-building and testing, is also regarded as essential by the growing number of environmental designers and managers who look to social science for a conceptual background to action.

The four approaches are illustrated in Figure 1,4. They range from a humanist or purist approach which seeks universals intuitively and neces-

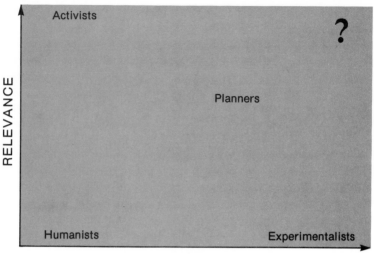

FIGURE 1,4 The four approaches

sarily eschews immediate relevance and scientific positivism, to the extremes of relevance characterized by environmental activists ("act now") and the extreme rigour of experimentalist social scientists ("before we can change the world we must first understand it"). Environmental designers and managers ("Planners," for short), confronting immediate issues and often having a fairly rigorous training, occupy an intermediate position. No group, as yet, has reached the top right corner of the diagram. Although this may be a consummation devoutly to be wished, in practice it is likely to be difficult to attain. Finally, it should be noted that the approaches outlined indicate tendencies only; they are not hermetic compartments. Individuals working in the field of environmental aesthetics frequently stray across the very permeable boundaries, and these interdisciplinary trespassers are often those most able to illuminate the common problems that confront all groups.

Humanists

Of the four approaches, the humanists tend to be the most contemplative. They are critical observers of human nature and landscape. Their concern is with the life of the mind, and with the contemplation of environments and human behaviour, rather than their manipulation. The approach is nonpositivist, idiographic, sometimes explicitly phenomenological or existentialist. Personal experience, intuition, and inductive reasoning

72

are stressed. In sharp contrast with mainstream scientific approaches, human values are at the forefront.

Tuan has emerged as one of the leaders of the field.[21] His works stress space and place as fundamental environmental components, and he repeatedly stresses the biological bases of human behaviour, the wide range of human experience, and the importance of culture. Both Tuan and Lowenthal have celebrated the cult of the past, nostalgia, and its influence on the aesthetics of landscape.[22] The movement for the preservation of old edifices may not be wholly based on aesthetics, but aesthetic arguments are frequently advanced as the chief reasons for preservation. Although our identities may in fact be bound up with a particular building or streetscape, appeals to architectural merit are more likely to carry weight with the authorities.

The humanists, indeed, have a strong leaning towards the past. Much emphasis is placed on tracing historic trends in landscape tastes. It is from the work of this group that we have come to appreciate the importance of learning and "currents of taste," such as the well-known fall and rise through time of Western appreciation of mountains and wilderness. The expressions of others, or consensual experience, is understood through the assessment of novels, poetry, paintings, diaries and other personalist art forms. This leads to a tendency to extrapolate the aesthetic feelings of the literate elite to society at large.[23] We know, for example, a great deal about the aesthetic theories of well-to-do eighteenth century Englishmen, and can still trace their expression in concrete form in field, park and town. We have very little comparable information about women or about the mass of the population who were compelled to live in and among these contrived landscapes. Only with the analysis of folklore or other plebeian art forms can any account be made of the tastes of the majority.[24]

Because of this rather elitist frame of reference, there has perhaps been a tendency to concentrate upon the exceptional in the landscape. Existing literature on environmental aesthetics tells us a great deal about Georgian cities, eighteenth-century country houses and estates, Italian hill towns, and French bastides, but very little about suburbs, apartments, and inner-city neighbourhoods, where the majority of our contemporaries live. In Relph's *Place and Placelessness* and Smith's *Syntax of Cities* the examples chosen to illustrate authenticity and high aesthetic value are almost universally pre-modern. In general, there seems to be an anti-modern, anti-urban bias, with a preference for architecture built under elitist or tyrannical forms of government.

Beyond history and culture, a significant trend among the humanists is to plumb the universals of experience in space and time. Important dichotomous concepts are space and place, home and abroad, inside and outside, front and back. The positivist construct of undifferentiated, objective space is contrasted with differentiated subjective place, pregnant with meanings, laden with emotions, and expressive of the intentions of its occupants. Similarly, the concept of home has been revived as the most valued environment beyond the individual's body. Hayward has investigated multidisciplinary approaches to the concept of home, Cooper sees house as the symbol of self, and Porteous has identified home as the territorial core.[25]

Much of the work produced by this group of landscape aestheticians is eminently readable. Indeed, the product of these investigations is as much a literary genre as a scholarly attempt to discover principles underlying our appreciation of landscape. It is notable also that a preference for the nonurban, and particularly for rural landscapes, is common among these practitioners. In many cases there is an acute lack of sympathy for the urban scene, with its lack of vegetation and water, and its suppression of seasonal rhythms. Some works are suffused with intimations of nostalgia for a Golden Age of rural life. Few humanists have felt able to turn their attentions fully towards the urban environments in which most of us live.

Activists

Activists are a varied group of academics, practitioners, writers and the general public, all of whom are concerned with issues of conservation, preservation, and rehabilitation of existing nonurban landscapes and urban fabrics. The will to conserve, which Tuan sees as having its basis in aesthetics, moral needs, and morale, is frequently fired by outraged feelings induced by what are regarded as unwarranted urban demolitions and nonurban despoliations.[26] Unlike the humanists, who are generally quietist in tone, the activists have immediately practical aims. In aesthetic terms, these aims are two-fold. First, activists wish to draw the attention of public and policy-makers alike to the growing degradation of both everyday and unique environments. This publicity campaign is vital, for:

> It is fairly obvious that many Americans really do not
> know that their land is blighted. We are so accustomed
> to environmental ugliness that we are surprised and of-
> ten indignant when the fact is called to our attention.[27]

The second aim is to find effective methods of curtailing this perceived despoliation, replacing it with policies designed to enhance the quality of the visual environment.

Time is of the essence. Environmentalists, conservationists, and eco-activists provide action, rhetoric, and scientifically-based arguments for the retention of aesthetic qualities in nonurban landscapes. Architects, designers, and other critics point to the growing ugliness of the urban scene and provide convincing arguments for the deceleration of planned change.

The literature produced is unashamedly emotional, and includes several luridly-illustrated volumes and at least one well-known film, McHarg's *Multiply and Subdue the Earth*. Angry architectural critics are particularly prominent; there is an aura of messianic fervour and apocalypse. Blake's classic *God's Own Junkyard* is a sickeningly-illustrated diatribe against "the planned destruction of America's landscape."[28] Ada Louise Huxtable, architectural critic for the *New York Times,* composes biting essays on the metropolitan landscape, some of which have been reprinted in the aptly-titled book *Will They Ever Finish Bruckner Boulevard?*[29] Ian Nairn has turned his clinical critical abilities against visual despoliation in Britain, the United States, and elsewhere.[30] Several humanist geographers have recently entered the fray, although Tuan and Lowenthal have been criticized as too meditative and scholarly to merit inclusion in the activist fraternity.[31]

Much support for practitioner activists comes from a groundswell of literary and popular opinion. Novelists and poets have railed against the aesthetic consequences of technology since the introduction of the canal during the English Industrial Revolution. More recently, George Orwell wrote scathingly of "the modern world, . . . everything sleek and streamlined, everything made out of something else. Celluloid, rubber, chromium steel everywhere, arc lamps blazing all night . . . no vegetation left, everything cemented over."[32] John Betjeman has specifically directed much of his poetry against the aesthetic depredations of modern planning. His "Planster's Vision" evokes a modern Britain in which old cottages are removed for the erection of "workers' flats in fields of soyabeans." In "The Town Clerk's Views" he mocks the planner's ideology of neatness and order:

> Hamlets which fail to pass the planner's test
> Will be demolished. We'll rebuild the rest
> To look like Welwyn mixed with Middle West:
> All fields we'll turn to sports grounds, lit at night
> From concrete standards with fluorescent light:
> And over all the land, instead of trees,
> Clean poles and wires will whisper in the breeze.[23]

Betjeman is also one of the few poets to celebrate the qualities of suburban landscapes.

One of the chief motivators of much environmental protest on aesthetic grounds is the feeling of impotence experienced by metropolitan citizens in the face of the threat of change. There appears to be a growing feeling against the egotistical nature of architects, and their predeliction for actualizing their fantasies by building self-serving Ozymandian monuments. The "primacy of the building" doctrine, based on the assertions of Corbusier and Mies Van der Rohe, has resulted in the conversion of landscapes into blandscapes, and sanctioned the efforts of the new Neros to raze the general heritage in favour of brutal monuments to corporate power and wealth. In a more general statement, expressing the powerlessness of the individual in the face of establishment control, Williams cries:

THE CITY IS EVIL. WE HAVE BEEN FOOLED.
WE HAVE BUILT Their PYRAMID.
IT IS Their GLORY. THE BUILDINGS ARE Theirs.
THE BIGGER THE BUILDINGS, THE BIGGER THE CITY,
THE BIGGER THE PYRAMID,THE SMALLER ARE WE.
they made us ants.
it was deliberate.
the bastards deserve to die.
but who the hell are they?[34]

The city is seen as a generator of accidie, alienation, anomie, and angst.

Effective citizens' action groups, however, have grown in importance in recent years. Aesthetic judgements are frequently an important component in citizen protests against planned change. Protests against urban renewal and the building of freeways have had conspicuous success in Boston, San Francisco, New York, New Orleans, and Washington, D.C.[35] In Victoria, British Columbia, during the early 1970s, there was a strong aesthetic component in the successful citizen movements to halt the construction of large waterfront buildings and to limit the height of downtown buildings to fourteen storeys.

It remains to be seen whether the activist movement will succeed with less spectacular items which have greater public acceptance, such as the billboarded, neon-lit, automobile-oriented strip which disfigures most entrances to North American cities and, in particular, belies Victoria's claim to be "a bit of Olde England." The automobile strip, while anathema to many aesthetic activists, is quite acceptable to most citizens and, indeed, in Smith's terms, provides the necessary "limbic logic" in urban scenes dominated by bland glass towers.[36] Further, many activists are much more articulate with regard to what is not liked than with what they would like to see. In this aspect, of course, aesthetic activism resembles many citizens' protest groups which band together mainly to oppose perceived threats.

Experimentalists

This group, largely composed of social scientists, lies at the opposite end of the relevance spectrum from the activists. The aims of the experimentalists include an attempt to tease out those landscape qualities or variables which are chiefly responsible for the generation of affect in the observer. It is also important to investigate the relationship between environment and individual in terms of the subject's attitudes, perceptions, and values. One of the chief contributions of the experimentalist approach is its operationalization of aesthetic concepts, particularly in terms of constructs such as satisfaction and preference.

Several assumptions underly this approach. Although it is possible, and most practicable, to have environmental assessments made by experts based on accumulated knowledge and experience coupled with the physical attributes of the site, the experimentalist's tendency has been to elucidate the salient components of the aesthetic feelings of nonexpert citizens.[37] A basic assumption in such environmental assessment research is that people judge the value of settings in terms of predefined environmental quality standards.[38] If such standards could be made explicit, contributions might be made towards the design of supportive environments.

Practically, much of this work has taken place under laboratory or quasi-experimental conditions. In consequence, little experimental work has involved subjects' assessments of real-world urban environments. In general, the environmental displays presented as stimuli involve some form of simulation. Such simulations may be static (a photograph) or dynamic (a film), abstract (a computer model) or concrete (a scale model).[39] Most research has used concrete, static displays, especially colour slides and photographs. Validation studies suggest that although observers' reactions to colour photographs are moderately predictive of their on-site responses, many other displays are of less value, and none compare with using the real environment as a stimulus.[40] Some attempt has therefore been made to use dynamic simulations. Films and videotapes have been brought into play, thus providing some of the environmental attributes absent from static displays. Sensations of smell and kinaesthesia, however, remain difficult to simulate.

The measurement of respondents' perceptions or assessments has generally been achieved through verbal rating scales such as the semantic differential. These are usually analyzed through inter-correlations between ratings, which are further reducible by factor analysis to a small number of dimensions. The problem here is that the result is purely descriptive; we gain little knowledge of the effects on affect of specific environmental characteristics. For instance, although it is common to find that nonurban scenes are usually rated as much more pleasant than urban ones, little information is available to specify which attributes, such as complexity, are salient in this assessment. As Wohlwill states, we learn little about "functional relationships among variables, which, one would suppose, is the business of a scientifically-based research effort."[41]

Clearly, while great attention has been given to technique development and the generation of empirically-derived models of preference, too little attention has been paid to the development of theory.[42] Wohlwill's extension of Berlyne's theory of aesthetics and Kaplan's landscape preference prediction work are exceptions to this trend.[43] Meanwhile, the less theoretical have expanded their range of assessment techniques to include a wide array of perceptual and behavioural techniques, including psychophysical scaling, behavioural mapping, and physiological monitoring procedures. Stokols suggests that little information exists on the relative validity of this battery of simulation and measurement procedures.[44]

Great care is usually taken in the sampling of respondents, although rather too many respondents turn out to be university students. The systematic sampling of various respondent groups has led to insights into the different perceptions and responses to environments of such groups as specialists and nonexperts, natives and visitors. In many cases, however, less attention has been paid to the choice of stimuli. Often the colour slides shown are rather arbitrarily chosen. Very few researchers have gone to the trouble to randomly sample the presented stimuli, although a notable exception is Milgram's study of New York City images.[45]

Despite these shortcomings, experimentalist work has been very fertile in confirming humanist concepts, as will be discussed below. The experimentalist position, however, generally involves a scientifically neutral attitude towards the stimulus. Activists and humanists, to varying degrees, frequently display negative attitudes towards modern cities. In contrast, the approach of the planner tends to have a more positive, value-laden, view of urban aesthetic quality.

Planners

Environmental designers and managers, less concerned with empirical research and theory-building than the social scientists, seek to create a harmonious blend of theory and applicability, with tendencies towards activism. The chief proponents are architects, city planners, and urbanologists.

Basing his work on Boulding's concept of the image, Lynch in 1960 propounded the notion of imageability, the ability of an environmental stimulus to evoke strong images in the mind of the observer.[46] An image is composed of identity, structure, and meaning. A five-element notation for analyzing image content was produced (paths, edges, nodes, landmarks, districts). Refinements of this approach are legion.

Image research has two chief goals. One is to enhance the city's legibility, to aid the citizen's sense of orientation, ease of movement, and wayfinding capacity. Equally important is the aesthetic goal. The *View from the Road* study, for example, aimed not only to make roads more legible and safe, but also more educative and pleasant to drive along.[47] This was to be achieved by the introduction of curves, signs, vegetation, landmarks, art objects, and the like, for Lynch was careful to stress that the goal of legibility could rapidly lead to boredom in North American grid-plan cities. A city richly supplied with landmarks, nodes, and deciduous vegetation would, through the operation of change in both space and time, enhance the opportunity for choice, satisfaction, and surprise.

Cullen's *Townscape* appeared almost simultaneously, complementing Lynch's American work with a British approach.[48] Like Lynch, Cullen's concern was with movement through the city and the changes such movement engenders in our appreciation of urban beauty. Cullen is far less concerned with legibility than with aesthetic satisfaction; to him "the city is a dramatic event in the environment," and the aim of the townscaper is to "manipulate the elements of the town so that an impact on the emotions [of the observer] is achieved."[49] He deals at length with serial vision, the sequence of views gained by turning corners, entering tunnels and alleys, and emerging from closure into courtyards and squares.

Townscape contains an immensely rich vocabulary for dealing with landscape qualities. Terminology, for example, includes the concepts of punctuation, rhythm, closure, exposure; congruity, complexity, mystery, surprise; hereness, thereness and place. Compared with Lynch, Cullen's vocabulary and analyses of urban environments are more three-dimensional, more microscale, and more evocative of aesthetic feeling. Yet the aims of the two schools are remarkably similar, namely to design towns from the point of view of the moving person, so that "the whole city becomes a plastic experience, a journey through pressures and vacuums, a sequence of exposures and enclosures, of contrast and relief."[50]

One of the chief problems here is the contrast between the needs of the moving automobile driver and the moving pedestrian. Cullen is concerned largely with pedestrians; Lynch admits freeways. In many Western cities pedestrian movement accounts for less than five percent of total outdoor travel.[51] Automobile travel precludes the perception of foreground detail, and nearby objects become blurred (Figure 2,4). At 40 mph, for example, roadside objects nearer than 40 feet become blurred, and loss of detail extends to about 80 feet from the car. Whereas the pedestrian view is concerned with microscale textures, changes of level, and surface detail, automobile occupants can only appreciate large-sized items. In automobile-

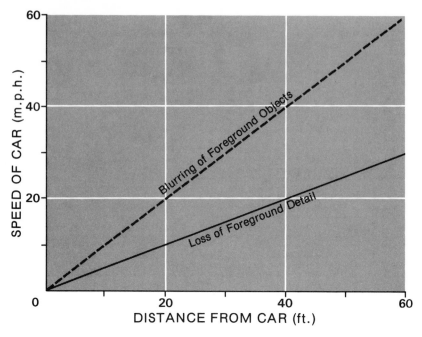

FIGURE 2,4 Automobile and sensory acuity

oriented cities, small high-maintenance vegetation types will be relegated to homesites and plazas, while major traffic arteries require large-sized, low-maintenance trees and bushes.

SYNTHESIS

The reader will have observed that the four approaches outlined above, far from being antagonistic, watertight compartments, are indeed complementary. Six dyad linkages exist, but these have clearly enjoyed very imbalanced development (Figure 3,4). Linkage between activists and experimentalists, for example, is much weaker than the interchange between activists and planners. Two linkages have been selected for discussion.

Humanist-Experimentalist Linkage

Humanists, as generators of basic concepts, clearly have links with all other approaches. In some recent cases, humanists have explicitly stated the value they expect their work to have for environmental designers and managers. Relph, for example, after a discussion of modern placelessness, calls for a new approach to urban planning "that is wholly self-conscious

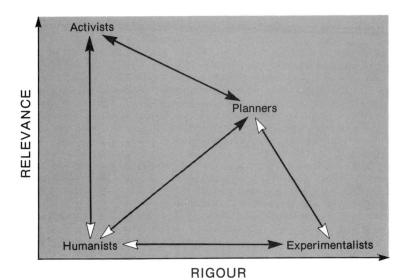

FIGURE 3,4 Linkages between the four approaches

. . . that is responsive to local structures of meaning and experience . . . that takes its inspiration from the existential significance of place . . ."[52] Tuan makes a similar attempt "to enter the debate on environmental design."[53] These are epilogue statements; the leap from desire to effective action, however, is difficult to make. In applicability terms, the chief benefit of the humanistic approach would seem to be the enhancement of the sensitivity to environment of both planners and public. In Tuan's words, the ultimate ambition is "to increase the burden of awareness."[54]

The linkage between humanist and experimentalist is more clearly documented. Several humanist notions of great vintage, and apparent in many cultures, are amenable to experimental verification. Such notions include the value of environmental complexity, the importance of mystery in the landscape, and the concept that human contact with nature has therapeutic value.[55]

Experimental work on the importance of the element of mystery in the landscape derives from Kaplan who, like Appleton, asserts that human survival necessitated that man not only be efficient in acquiring landscape information, but also that he gain satisfaction by processing such data.[56] If *Homo sapiens* likes to acquire environmental information, an environment containing elements of mystery will be especially preferred because it suggests to the viewer the possibility that additional information is discoverable. The environmental aesthetician's familiar curving path is a prime example of mystery which encourages exploration (Figure 4,4). In a cross-cultural study using black-and-white photographs of roadside scenes it was dis-

FIGURE 4,4   Curving paths, illustrating mystery

covered that scenes involving mystery were highly preferred.[57] Subjects also favoured scenes with even ground textures, medium to high levels of depth, and the presence of a focal point. Complexity, surprisingly, was not favoured, owing, it is thought, to the experiment's inability to distinguish between random and patterned complexity.

In relation to the "nature tranquillity hypothesis," which *inter alia* is responsible for the prevalence of urban parks, an experiment was devised in which colour slides of outdoor environments were shown to mildly-stressed subjects.[58] Both urban and nonurban scenes were used, and the reactions of subjects scored via standard tests. Exposure to nonurban scenes significantly reduced fear arousal among the subjects, while feelings of affection, friendliness, playfulness, and elation were increased. Urban scenes, however, had generally the opposite effect. In particular, viewing the urban scenes significantly increased feelings of sadness.

Such research provisionally confirms intuitive feelings concerning the therapeutic value of nature and the anxiety-laden atmosphere of cities. Several implications are apparent. First, deeper investigation is required into the specific components of urban and nonurban scenes which trigger psychological feelings. Second, a physiological basis for aesthetic feelings is suggested. Third, should such research be validated, a considerable and valuable input into urban planning theory is likely. There is already considerable agreement between experimentalists and planners on the value of the environmental attributes of novelty, surprise, ambiguity, coherence, clarity, and complexity. As yet, however, a wide applicability gap remains between experimentalist and planner.

Activist-Planner Linkage

The gap between planner and activist can more readily be closed, for in both the tendency towards relevance usually triumphs over desires for rigour. A single case will be taken as exemplifying the potential for collaboration between planner and citizen.

During the last two decades San Francisco's character has undergone great transformation through the proliferation of highrise structures. Many Bay Area residents protested, fearing that a city known for its taste, charm, and views of hills and bay would lose its visual personality. To their support came design professionals who were able to provide telling analyses of the visual problems created by tall buildings.[60] These problems include shape, colour, tone, scale, street effects, view blockage, and symbolic meaning. For example, dark brownish-black, hard-edged, reflecting glass towers sit rather uncomfortably in a soft, creamy, light-coloured, low-rise context.

By providing vocabulary and measurement techniques, planners were able to act in collaborative, advisory, and advocacy roles vis-a-vis nonexperts. At this stage in planning development the interventions of inperts frequently require such assistance.[61]

Further, by use of simulation techniques it is possible to demonstrate to public and policymakers alike the likely aesthetic impact of proposed developments or alternative futures. The Berkeley Environmental Simulator, for example, provides a three-dimensional scale model of San Francisco. The proposed development may be placed in context, and then viewed by means of a travelling camera which permits views from several angles, and from the point of view of both the pedestrian and the automobile driver. Appleyard suggests that such simulations should be readily available to the public before future developments are discussed; he speaks of "truth in simulation."[62] Sample surveys of San Francisco citizens suggest that unless people's aesthetic objectives are met or catered to, satisfaction with San Francisco as a visually-pleasing environment will significantly decline with the growth of tall buildings.[63]

Other linkages are clearly in operation, and in particular the experimentalists are keen to put forward their techniques as practical inputs into the aesthetic planning of urban environments.[64] It is only through the mutual collaboration of all four groups that intuitive concepts will be verified, techniques for aesthetic assessment be developed and incorporated into the planning process, and action to enhance the aesthetic quality of cities be organized. Divided, the four approaches will achieve significantly less.

THE PROBLEM OF SCOPE

The interdisciplinary field of environmental aesthetics is excessively concerned with the visual quality of landscape. Both lack of work on the other senses and experiments comparing several senses suggest that vision is indeed dominant in human perception.[65] Yet environmental experience, aesthetic or otherwise, is holistic experience. Paintings of townscapes are indeed visual experiences, but a walk through a landscape garden or a crowded city is a total sensory experience—visual, auditory, olfactory, tactile, kinaesthetic; and in some cities one can taste the air.

Urban legislation suggests that citizens may be equally concerned with their auditory and olfactory environments as with visual amenity. Zoning regulation with regard to odoriferous "nuisances" stresses this point. I therefore urge the widening of our notion of environmental aesthetics to include all aspects of sensory experience. This involves new research frontiers and perhaps new vocabularies. Brief examples follow.

The Auditory Environment

With the notable exception of Southworth's work on the auditory regions of Boston, and the more sustained research of Schafer in Canada, very little has been done to elucidate the aesthetic quality of the aural environment.[66] Schafer and his associates, working on the World Soundscape Project, were compelled to invent their own terminology. Using the analogy of the visual landscape, they suggest such terms as soundscape, soundmark, and earwitness.

The relative ease of aural environment surveys derives from the possibility of producing both objective, measurable data (via a sound-meter) and subjective data via interviews and questionnaires. Sounds, furthermore, are recordable and thus may be used in environmental displays in much the same way as aesthetic experimentalists now use photographs and films.

Aural research is still in its infancy, but has already produced some standard descriptive tools such as the isobel map, the sound event map, and the "listening walk" map.[67] In the latter a Lynch-type walk around the block produces a map of sound sources, intensities, durations, observer affect, and the like. It is reasonably easy to make comparisons of sound events and soundscapes through both time and space.

The Olfactory Environment

The ease of recording, measurement, and analysis which characterizes soundscape studies is not possible for the olfactory environment. Again, a new vocabulary is needed, involving such terms as odourscape, odourmark, and nosewitness. I am not presently aware of any studies of olfactory environmental aesthetics, although many statements have been made on the tactile, gustatory, and olfactory value of such elements as water.

One pilot study by my own students relied for measurement on the assessments of a number of individuals who had learned to recognize typical smells organized according to the Zwaardemaker odour classification.[68] Odiferous substances typical of Zwaardemaker's nine classes of smell were used to train this expert panel.

The panel was then asked to identify the predominant odours at specific locations and rate their preference for these on a five-point scale. Considerable agreement among panel members suggests some validity for this approach. Pet shops and the rear entrances of supermarket meat departments were universally classed as Zwaardemaker IX (nauseating) and rated very low on the preference scale. Most panel members were able to identify a tea and coffee store as Zwaardemaker VI (empyreumatic). The potential here for enhancing the environmental sensitivity of the individual, as well as for environmental amelioration, seems worth investigating.

The Tactile-Kinaesthetic Environment

These responses are also difficult to monitor effectively. Again, data must be gathered largely from the immediate subjective experience of the respondent. Although a considerable amount of research has focussed upon the experience of moving through the city, the emphasis has been almost wholly upon visual quality. Few attempts have been made to consider the sense of touch or feelings of motion or change in orientation. These sensory experiences, like taste and smell, are intimately related, and are in turn strongly related to the visual experience.

Moreover, there are both intrinsically pleasurable and practically valuable applications. Regarding the former, one thinks of the Japanese urban garden, where one is constantly encouraged to change speed and direction, feel different textures underfoot, and appreciate both sharp angles and curves. Changes in view are stressed by the necessary body shifts involved in crossing a line of stepping-stones (Figure 5,4). Such principles could be used to reduce the banality of many indoor shopping malls. On city streets, tactile cues include warning grooves or ridges in the road surface, where the tactile sensation is strengthened by auditory and visual cues such as noise, signs and lights.

It is unlikely that nonvisual environmental aesthetic research will ever have high priority. Yet nonvisual sensory research could be extremely useful in the enhancement of the navigability of blind and physically handicapped persons.

A CASE FOR ENVIRONMENTAL AESTHETICS

It is rather easy to develop a case against environmental aesthetics as a priority in either research or practice. Aesthetic satisfaction is not high in the hierarchy of human needs. Its importance varies greatly between individuals, culture groups, and classes. A number of studies indicate that strong environmental aesthetic concern is limited to certain social classes. Porteous's work on northern British Columbia towns suggests that the aesthetic dimension is of little importance to the northerner.[69] British research suggests that middle class people attach greater significance to townscape features of high architectural quality or historic interest than do their working class counterparts.[70] In his assessment of the environmentalist movement, O'Riordan comes to the same conclusion.[71] Comparing urban planners and nonprofessional citizens in Cuidad Guyana, Venezuela, Appleyard found that the latter had a much lower, often negligible, appreciation of building quality or symbolism.[72] It seems inescapable that whereas the majority of the population is still concerned with "standard of living," only a minority can afford to consider "quality of life."

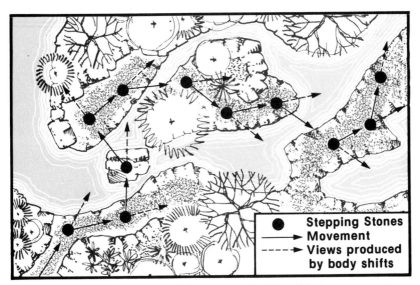

FIGURE 5,4 Stepping-stones: the link between visual and kinaesthetic sensory experience

Relph's essay on place and placelessness exemplifies this, perhaps inevitable, aesthetic elitism.[73] The comments of thinkers from de Toqueville and Ruskin to Cullen are marshalled to point out the erosion of pleasing, placeful urban environments in favour of "chromium-plated chaos." Relph judges as misguided: Baedeker tourists who check off starred landmarks; families who consume food, but do not dine, in Burgerland; dwellers in endless suburbia; travellers on billboarded freeways; and purchasers of ubiquitous kitsch. Yet no evidence is given to demonstrate that these intellectual views are common throughout social classes, life-cycle stages, or life-styles. In Smith's terms, indeed, modern blandscapes are redeemed by the neon Burgerland strip, and Venturi claims that we can learn from Las Vegas.[74] J.B. Jackson stresses the value of such other-directed townscapes.[75] Further, despite the cries of the critics and the grandiose plans of Le Corbusier, most North Americans and Australians, together with many Europeans, have voted with their feet for the conveniences of a suburban Broadacre City.

Unfortunately, the case for environmental aesthetics also has elitist overtones. The objectives of environmental planning have traditionally been separated into the three sometimes contradictory goals of the protection of physical and mental health, the enhancement of economic value, and the preservation of beauty.[76] These objectives have generally been set by policy makers and planners. Recent citizen involvement in aesthetic decisions, however, suggests that public participation may become an important force as aesthetic concepts and vocabularies filter through societal strata. Environmental education will likely enhance citizen sensitivity to urban environments. And if the experimental support for humanistic ideas of city-induced anxiety and nature-generated tranquillity is validated, environmental aesthetic research will become increasingly important in a neurotic posturban world.

Urban environmental aesthetics, however, has not yet come of age as a discipline. Only recently has it become acceptable even to use the term "aesthetic" in the urban environmental context. Indeed, even in nonurban evaluations, "aesthetics" is generally eschewed in favour of euphemisms or specific dimensions such as preference, perception, or connotation. According to Heath, there is strong prejudice against scientific investigation of aesthetics in the architectural context, and this is the first hurdle to overcome.[77] Heath's title offers an instructive epilogue: "Should We Tell the Children About Aesthetics, or Should We Let Them Find Out in the Street?"

REFERENCES

1. MASLOW, A., *Motivation and Personality*. New York: Harper & Row, 1974.

2. PAWLEY, M., *The Private Future: Causes and Consequences of Community Collapse in The West*. London: Thames & Hudson, 1974.

3. LYNCH, K., *Managing the Sense of a Region*. Cambridge, Mass.: The MIT Press, 1976.

4. LOWENTHAL, D., *Finding Valued Landscapes*. Toronto: University of Toronto, Evironmental Perception Research Working Paper No. 4, 1978.

5. PENNING-ROWSELL, E.C., *Alternative Approaches to Landscape Appraisal and Evaluation*. Enfield, Middlesex: Middlesex Polytechnic, Planning Research Group Report No. 11, 1973; WILSON-HODGES, C., *The Measurement of Landscape Aesthetics*. Toronto: University of Toronto, Environmental Perception Research Working Paper No. 2, 1978.

6. WHITE, M. and WHITE, L., *The Intellectual Versus the City*. Cambridge, Mass.: Harvard University Press, 1962.

7. DAVIS, K., *Cities: Their Origin, Growth and Human Impact*. San Francisco: W.H. Freeman, 1973.

8. PORTEOUS, J.D., *Environment & Behavior: Planning and Everyday Urban Life*. Reading, Mass.: Addison-Wesley, 1977, pp.2-5.

9. DOWNS, R.M., "Cognitive Structure of an Urban Shopping Centre", *Environment and Behavior*, 2, 1970, pp. 13-39; ULRICH, R., *Scenery and the Shopping Trip: The Roadside Environment as a Factor in Route Choice*. Ann Arbor, Michigan: Michigan Geographical Publication No. 12, 1974.

10. APPLETON, J., *The Experience of Landscape*. London: Wiley, 1975; SMITH, P.F., *The Syntax of Cities*. London: Hutchinson, 1977.

11. APPLETON, *op. cit.*, p. viii.

12. *Ibid.*, p. 70.

13. *Ibid.*, p. 73.

14. *Ibid.*, p. ix.

15. *Ibid.*, pp. 194-202.

16. *Ibid.*, p. 195.

17. SMITH, *op. cit.*

18. *Ibid.*, p. 214.

19. *Ibid.*, p. 106.

20. ROEPKE, H.G., "Applied Geography: Should We, Must We, Can We?", *Geographical Review*, 67, 1977, pp. 481-82.

21. TUAN, Y-F., *Topophilia*. Englewood Cliffs, N.J.: Prentice-Hall, 1974.

22. TUAN, Y-F., *Space and Place: The Perspective of Experience*. Minneapolis: University of Minnesota Press, 1977, pp. 179-198; LOWENTHAL, D., "Past Time, Present Place: Landscape and Memory", *Geographical Review*, 65, 1975, pp. 1-36.

23. LOWENTHAL, D. and PRINCE, H.C., "English Landscape Tastes", *Geographical Review*, 55, 1965, pp. 186-222.

24. BUNKSE, E.V., "Commoner Attitudes Toward Landscape and Nature", *Annals, Association of American Geographers*, 68, 1978, pp.551-66.

25. COOPER, C., "House as Symbol of Self," in LANG, J., *et al.* (eds.), *Designing for Human Behavior*. Stroudsburg, Pa.: Dowden, Hutchinson and Ross, 1974, pp. 130-146; HAYWARD, D.G., "Home as an Environmental and Psychological Concept", *Landscape*, 20, 1975, pp. 2-9; PORTEOUS, J.D., "Home: The Territorial Core", *Geographical Review*, 66, 1976, pp. 383-90.

26. TUAN, *op. cit.*, (reference 22), p. 197.

27. LEWIS, P.F., "The Geographer as Landscape Critic", in LEWIS, P.F., *et al.*, *Visual Blight in America*. Washington, D.C.: Association of America Geographers, Resource Paper No. 23, 1973, p. 2.

28. BLAKE, P., *God's Own Junkyard*. New York: Holt, Rinehart & Winston, 1964.

29. HUXTABLE, A.L., *Will They Ever Finish Bruckner Boulevard?* New York: Macmillan, 1970; CLIFF, U., "New York's Better Self", *Design and Environment*, 2, 1971, p. 2.

30. NAIRN, I., *The American Landscape: A Critical View*. New York: Random House, 1965.

31. MEINIG, D.W., "Commentary: Visual Blight - Academic Neglect", in LEWIS, P.F., *et. al., op. cit.*, p. 45.

32. ORWELL, G., *Coming Up for Air*. London: Secker & Warburg, 1939.

33. BETJEMAN, J., *Collected Poems*, Enlarged Third Edition. London: Murray, 1970.

34. WILLIAMS, G., *The Man Who Had Power Over Women*. New York: Stein and Day, 1967, p. 142.

35. PORTEOUS, *op. cit.*, (reference 8), p. 327.

36. SMITH, *op. cit.*

37. ROGERS, A., "Landscape Evaluation in Planning Rural Areas", *Area,* 11, 1979, p. 10.

38. STOKOLS, D., "Environmental Psychology", *Annual Review of Psychology*, 20, 1978, pp. 253-95.

39. MCKECHNIE, G.E., cited in STOKOLS, *op. cit.*

40. STOKOLS, *op. cit.*, p. 267; SEATON, R.W. and COLLINS, J.B., "Validity and Reliability of Ratings of Simulated Buildings", in MITCHELL, W.J., (ed.), *Environmental Design: Research and Practice.* Los Angeles: EDRA 3 Proceedings, 1972, pp. 6.10.1-12.

41. WOHLWILL, J.F., "Environmental Aesthetics: The Environment as a Source of Affect", in ALTMAN, I. and WOHLWILL, J.F., (eds.), *Human Behavior and Environment: Advances in Theory and Research, Vol. 1.* New York: Plenum Press, 1976, p. 61.

42. WOHLWILL, *op. cit.*; STOKOLS, *op. cit.*

43. KAPLAN, S., "Adaptation, Structure, and Knowledge", in MOORE, G.T. and GOLLEDGE, R.G., (eds.), *Environmental Knowing.* Stroudsburg, Pa.: Dowden, Hutchinson and Ross, 1976, pp. 32-45; WOHLWILL, *op. cit.*

44. STOKOLS, *op. cit.*

45. MILGRAM, S., "Psychological Map of New York City", *American Scientist*, 60, 1972, pp. 181-194.

46. LYNCH, K., *The Image of the City.* Cambridge, Mass.: The MIT Press, 1970.

47. APPLEYARD, D., *et al.*, *The View From the Road.* Cambridge, Mass.: The MIT Press, 1964.

48. CULLEN, G., *Townscape.* London: Architectural Press, 1961.

49. CULLEN, G., *The Concise Townscape.* London: Architectural Press, 1971, p. 9.

50. *Ibid.*, p. 9.

51. ULRICH, R., "Urbanization and Garden Aesthetics", *Longwood Seminars*, 8, 1976, pp. 4-8.

52. RELPH, E., *Place and Placelessness.* London: Pion, 1976, p. 146.

53. TUAN, *op. cit.*, (reference 22), p. 202.

54. *Ibid.*, p. 203.

55. RYBACK, R.S. and YAW, L., "The Magic of Water", *Man-Environment Systems*, 6, 1976, pp. 81-83.

56. KAPLAN, S., "The Dimensions of the Visual Environment", in MITCHELL, W.J., (ed.), *Environmental Design: Research and Practice*. Los Angles: EDRA 3 Proceedings, 1972, pp. 6.7.1-5; and "Cognitive Maps, Human Needs, and the Designed Environment", in PREISER, W., (ed.), *Environmental Design Research*. Stroudsburg, Pa.: Dowden, Hutchinson and Ross, 1973, pp. 275-283.

57. ULRICH, R.S., "Visual Landscape Preference: A Model and Application", *Man-Environment Systems*, 7, 1977, pp. 279-93.

58. ULRICH, R.S., "Visual Landscapes and Psychological Well-being", paper presented at the Ninth Meeting of the Environment Design Research Association, Tucson, Arizona, April, 1978.

59. PORTEOUS, *op. cit.*, (reference 8), pp. 226-232; RAPOPORT, A. and KANTOR, R.E., "Complexity and Ambiguity in Environmental Design", *Journal of the American Institute of Planners*, 33, 1967, pp. 210-21; VENTURI, R., *Complexity and Contradiction in Architecture*. New York: Museum of Modern Art, 1966.

60. APPLEYARD, D. and FISHMAN, L., "High-Rise Buildings Versus San Francisco", in CONWAY, D., (ed.), *Human Response to Tall Buildings*. Stroudsburg, Pa.: Dowden, Hutchinson and Ross, 1977, pp. 81-100.

61. PORTEOUS, J.D., "Design With People: The Quality of the Urban Environment", *Environment and Behavior*, 3, 1971, pp. 155-178.

62. APPLEYARD, *op. cit.*

63. DORNBUSCH, D.M. and GELB, P.M., "High-Rise Visual Impact", in CONWAY, *op. cit.*, pp. 101-111.

64. STOKOLS, *op. cit.*, p. 268.

65. VICTOR, J. and ROCK, I., "Vision and Touch: Experimentally Created Conflict between the Two Senses", *Science*, 143, 1964, pp. 594-96.

66. SOUTHWORTH, M., "The Sonic Environment of Cities", *Environment and Behavior*, 1, 1969, pp. 49-70; SCHAFER, R.M., *The Tuning of the World*. Toronto: McClelland and Stewart, 1977.

67. SCHAFER, *op. cit.*, p. 264-270.

68. HALL, S.I. and COLEMAN, B.H., *Diseases of the Nose, Throat and Ear*. London: Livingstone, 1973; STELL, P.M. et al., *Ear, Nose, and Throat*. London: English University Press, 1974.

69. PORTEOUS, J.D., "Quality of Life in British Columbia Company Towns: Residents' Attitudes", in PRESSMAN, N., (ed.), *New Communities in Canada: Exploring Planned Environments*. Waterloo, Ontario: University of Waterloo, 1976, pp. 332-46.

70. GOODCHILD, B., "Class Differences in Environmental Perception", *Urban Studies*, 11, 1974, pp. 157-69.

71. O'RIORDAN, T., *Environmentalism*. London: Pion, 1976.

72. APPLEYARD, D., "Why Buildings are Known: A Predictive Tool for Architects and Planners", *Environment and Behaviour*, 1, 1960, pp. 131-56.

73. RELPH, *op. cit.*

74. VENTURI, E. and BROWN, D.S., *Learning from Las Vegas*. Cambridge, Mass.: The MIT Press, 1972.

75. JACKSON, J.B., "Other-directed Houses", *Landscape*, 6, 1956, pp. 29-35.

76. KATES, R.W., "Comprehensive Environmental Planning", in HUF-SCHMIDT, M.M., (ed.), *Regional Planning.* New York: Praeger, 1969, pp. 67-87.

77. HEATH, T.F., "Should We Tell the Children about Aesthetics, or Should We Let Them Find Out in the Street?", in CANTER, D. and LEE, T., (eds.), *Psychology and the Built Environment.* London: R.I.B.A., 1974, pp. 179-183.

PLATE 6   Landscape as Scenery: Canyon Country of the Yellowstone.   *B. Sadler Photo* ▶

5 VISUAL ASSESSMENT OF NATURAL LANDSCAPES

R. Burton Litton, Jr.
University of California, Berkeley

INTRODUCTION

Aesthetic dimensions of natural landscapes are complex and elusive. They need to be abstracted to the degree possible and made more tangible if they are to become useful agents in wildlands resource management. In their application to landscape planning and resource management, they are being interpreted as visual dimensions.[1] This visual shortcut still carries a hidden agenda—a baggage of aesthetics; we should appreciate this complexity and eventually expect to treat the problem in more adequate fashion. Santayana says that the landscape is an "indeterminate object," allowing the observer considerable latitude in composing elements and relationships so that there is aesthetic response—an emotional judgement, a gut reaction, that quality is present.[2] Whether we refer to the quality of the landscape as visual, scenic, or aesthetic, it is essential to consider it as a part of environmental quality. In this way the landscape can be seen as a resource in its own right, with its visual integrity being connected to both natural processes and sensitive land management.

ELEMENTS OF AESTHETIC EXPERIENCE

There is a broad set of implications about people's aesthetic experience and the landscape. These are overly simplified (or ignored) in wildlands management applications. The explanation for omission is simple enough. Research is needed; no effective or comprehensive way of gathering such insight is now available. Prospects for much agreement are poor.

Three environmental aspects of aesthetic experience are identified.[3] There is the observer's state of mind—his attention or inattention to the passing landscape. The driving salesman is thinking about his next sales contact rather than his surroundings. There is the context of observation— is the observer in the process of delivering a D-8 Caterpillar tractor to a

97

worksite, or is he/she on vacation, sitting on a rock listening to the sound of the Bow River? Then there is the landscape itself—the "environmental stimulus"; is it the Canadian Rockies north of Banff or is it the prairie country east of Edmonton?

PROFESSIONAL LANDSCAPE ANALYSIS

The individual professional planner, landscape architect, or environmental planner, in relation to resource management, is also called upon to document the landscape as a resource by using visual analysis. But what does this mean? Landscape analysis is a catch-all term. It can be a matter of who defines it—what individual, what agency. Landscape analyses should be understood to include physical—visual (or scenic) inventories, landscape assessments to relate to environmental analysis, predictions of visual impact for alternative development proposals, and the codification of visual controls for landscape planning and design. Additionally visual analysis is part of a research field with diverse, intermingled dimensions— aesthetic, design, perceptual, socio-psychological, silvicultural, ecological and others. It is commonly and erroneously suggested that scenic evaluation (or assessment) for resource planning is the only concern of visual analysis. While this is important, it is a narrow view. To avoid confusion, the varied facets of visual analysis call for clear definition.

Then there are questions. Will the professional have sufficient training and experience to do an adequate job of landscape inventory and subsequent evaluation? It takes time to know a landscape. Time and familiarity with a region or locality are necessary ingredients if a professional is to do a proper job. The landscape is also dynamic and changing, its image differing because of diurnal change, the seasons, and the much longer time spans of natural processes. It is also essential that the professional go into the field to make visual landscape studies.[4] Some preliminary guidance may be obtained from topographic maps, from reading geographic, vegetation or historic reports and accounts. But these are preliminary preparations after which there is no substitute for field observations and related recordings.

A central issue of professional landscape analysis is whether or not objectivity is possible or in what measure. My work as an individual landscape architect is apt to be looked upon as subjective, reflective of personal values and constrained by my own knowledge and experience. Yet I consider my work as based upon a logical application of my field of expertise, a defence against that which may merely be called "intuitive." Subjectivity of a professional does occur when clients or an audience are left uninformed as to

98

what criteria are employed or when esoteric procedures appear to be intuitive because explanation is omitted. Anyone who works in the hazy world of landscape aesthetics inevitably grapples with subjective problems, whether perceptual values and opinions of a lay public or those of a single professional person.

Complete avoidance of personal bias in scenic analysis is probably not possible, but it is a clear goal to seek. There are a number of ways to approach this. Straightforward descriptive representation of landscape, carefully observed, should first of all provide base line information about existing conditions. To document what is present does not depend upon judgements of value but rather upon faithfulness to recording visual elements and relationships. Abstraction and omission of certain detail is necessary justification still depending upon criteria being applied. If professional criteria can be sympathetic to known public perceptions, they gain credibility. Knowledge of historic and cultural backgrounds about a particular region should be represented in professional approaches; this may be the only available insight about certain people's sense of landscape values. Another charge to the professional is to account for normal seasonal changes of region and place, to anticipate where and when landscape images will be more distinctive or less. None of us can be familiar with all regions and their specific places, but we can work with people who are.

Lastly, emphasis upon objectivity has to do with appreciating the landscape's complexity and that it consists of both tangible and intangible forms. It exists as a set of physical parts and relationships seen in the round—my primary image as a landscape architect, the image altered by resource manipulations. It exists as a perceptual composition in the mind, a memory, and anticiapation—a psychologist's image. It exists as a gut feeling—which may be anyone's image. The landscape is all of these and more—and all simultaneously.

LANDSCAPE INVENTORIES AND EVALUATIONS

Landscape inventories record physical-visual elements and relationships observed at a particular point in time. Basic inventory elements and relationships consist of landforms, vegetation mosaics, and water bodies described through such visual design terms as forms, space, scale, color, pattern, and compositional type.[5] In wild landscapes, land use patterns are also part of the visual inventory. They may presumably be a minor element of description unless they are degrading or disruptive due to such causes as chaotic relationships, overbearing scale or overt contrast. Grant Jones uses the criterion of *intactness* to describe the relative degree of apparent naturalness.[6]

Visual inventories are of two types—routed and areal. The routed inventory uses a road, trail or stream as the location of a traveling observer with attention limited to the landscape within the visual corridor.[7] The area seen from the road does include invisible pockets. In lands with even modest relief, the visual corridor of a routed inventory tends to be relatively restricted, indicating that there is (or can be) intensive attention to detail at small scale. Areal inventories may be of any scale, of large regional landscapes or of small places.[8] Their content may be adjusted to suit either broad planning purposes or those of intensive design.

Media for the visual inventory include mapping, narrative description, and representative graphic samples—photos, diagrams, sketches.[9] These are primary and conventional tools both for field work and finished reports. Computer products and video also have application.

Topographic maps are fundamental for fieldwork and recording. For routed inventories, maps at 1:20,000 - 1:24,000 and 1:50,000 - 1:62,500 scales are most appropriate. Larger scale maps, 1:100,000 - 1:250,000 scales, are desirable to give a sense of context for those of smaller scale and are also well suited for broader, more generalized area inventories. Prefield topographic map study can provide some clues about the nature and continuity of typically repetitive land forms, discontinuities or anomalies suggested by breaks or contour pattern, the presence of water elements and an identification of stream patterns. Gross patterns of vegetation in relation to topography and streams are suggested by plant cover overprinting; there are normally hints present about land use patterns, urbanization and other man-made modifications.

A routed inventory can serve as an example in suggesting what a landscape inventory includes and how it may be approached in the field. Using a route, whether a road or a wild river, the visual corridor establishes boundaries and there is an automatic choice made of a linear platform for observation. This is simpler than an area inventory because of sheer size and also avoids questions of how best to do field reconnaissance. For any study area, whether identified as regional or local, a primary principle for the inventory is to record *typical* expressions of landform, vegetation, water elements, and land use patterns along with the atypical—the rare or unusual. This recording should account for the whole visual continuum consisting of ordinary landscapes and those which are unusual or extraordinary (Figure 1,5).

Landforms are the skeletal backbone of the landscape. I have suggested that landforms as visually perceived be thought of as convex upright forms and as concave spaces—positive and negative form-space concepts.[10] Accordingly, landforms for visual inventories can be characterized as to their descriptive contours (skyline silhouettes), whether isolated or repetitive elements of a range (a local community), their scale (relative sizes), or their surface variance (color or vegetative overlay).[11] Locally typical skylines—the form of hills or mountains and their relative relief may be suggested by diagrams such as (see Figure 2,5):

Spaces in the landscape are valleys, canyons, or swales. For inventories they are better described as enclosed landscapes with their visual dimensions identified (Figure 3,5). Proportions of walls to floor (vertical facade to bottomland), scale, configuration of floor plans (simple to complex), and the materials of walls compared to floors (barren cliffs to riparian vegetative cover, for example) suggest what enclosed spaces look like. Diagram sections can compare proportions such as:

Vegetative mosaics constitute the main patterns of surface variance upon the landscape. These are seen in two ways; they are broad overviews observed from a distance and intimate views seen close at hand.[12] Some record of both these attributes should enter the inventory. Different life forms of vegetation, their color-texture-scale contrasts, their interlocking pattern shapes, edges, and dispositions, along with seasonal changes, make up basic elements and relationships. Such visual relationships are interpretations made from observed forests and woods (coniferous and broadleaf trees), riparian aggregations, grassland, scrub or chaparral, and also barren surfaces—rock or mineral surfaces devoid of vegetation. The inventory may represent different plant types according to proportioned coverages for a particular corridor face but also needs to include descriptive and graphic representations of vegetation patterns (Figure 4,5).

Water is a special and rare element in the landscape. Or it is rarely as commonplace as dry land. It has a life of its own, a life both aesthetic and biotic. Its response to diurnal change, to weather and season makes it more akin to sky and atmosphere than to land. Yet as a primary agent in the making of landforms, there is always a native fit between water and surrounding landscape. There is also the visual connection between water and riparian vegetation, often an addition of more bright and vivid colors compared to the darker, more somber colors of upland forest. Water, then, fills numerous roles in the landscape; it can be a feature in itself (Figure 5,5); it is visually integrative with vegetation and enclosing landforms.[13] For the inventory, stream paths are classified as straight, sinuous, braided or meandering.[14] Water's movement needs to be noted—fast to slow, rapid white to still dark. Lakes can be accounted for by their sizes, disposition, and especially their complexity (or simplicity) of configuration and edges. Presence of riparian vegetation is visually significant, sometimes masking water but at the same time making its presence known. Then there is need to take notice of the forms of land, the enclosure closely associated with water, varying from flats to slopes to cliffs (Figure 6,5).

Land use elements within wild landscapes may be seen as fragments or segments of man-made change; but if an area of sufficient size is inventoried, these same things may be repeated into characteristic patterns. Such patterns can be those of agriculture, forestry, mining, recreation, road and transmission corridors. The inventory first documents what is found, whether fragments or patterns. Instant assumption that all land use elements are visual intrusions upon the surroundings is an error. The record may be instrumental in determining that land use patterns have positive values as indigenous, historical parts of landscape. Visual analysis principles are the same whether land use elements are seen as compatible with or degrading of surroundings. Scale, color, geometric form, locations with respect to sky, water, and vegetation edges, relative density of patterns, the handling of every-present roads[15]—these and other design based analyses can abstract land use characteristics to be entered in the inventory (Figure 7,5).

Any visual inventory becomes a better tool for evaluation and later management application if it is divided into units.[16] Each unit represents discernible variations within the landscape study area. Definitions of units depend upon spatial characteristics and relationships of landforms, vegetation and water patterns or upon presence of a set of visually consistent (or homogenous) elements.[17] Units help overcome the generic problem in landscape assessment of comparing pyramids with horse troughs, pine trees with lakes. Units which are distinguished by dominant water bodies and patterns may be compared with like units. Those with general similarities of significant landform features can be contrasted. A unit can also represent rarity.

It has been my conviction that scenic resources should be evaluated by the use of aesthetic criteria. There are three of them: *units*, *variety*, and *vividness.*[18] It is of importance to recognize that these three measures of quality apply simultaneously and that they are in a potential state of conflict with one another.[19] Because these criteria are normally applied to works of art, their application to landscape calls for interpretation.[20] It is also apparent that unity, variety, and vividness are abstract concepts to persons unfamiliar with them. If they are to serve an evaluation purpose in landscape planning, they need to assume an air of practicality.

Unity is the quality of all parts being joined together into a single and harmonious whole. In the landscape, one representation of it is in visual units. Unity is also expressed by landscape compositional types, one of which is a feature dominated landscape.[21] An example is that of an isolated mountain peak, large in scale and of an unusual skyline contour, dominating a set of subordinate, smaller scale peaks and ridges along with their forest cover, their lakes and streams. Monotony is also unity but of low quality with variety and vividness in scant supply. But monotony can also be the observer's incapacity to understand or to take into account a subtle level of richness and vividness.

Variety is the number of different elements and relationships that may be found. Diversity or richness carries the same idea—it is just as important to the biologist as to the artist. Greater variety is equated to higher quality, but there is the need for connective composition (unification) to thwart potential visual chaos. In the landscape, variety can be the presence of deciduous hardwoods among conifers, a stream of water as additional to upland forested slopes. The linear or trellis pattern of riparian stringers of aspen along a creek system, set in a matrix of conifers, represents a simple version of composed variety.

Vividness is, most simply, the presence of contrasting things seen together. While it is related to variety, it is not the same as the mere contrast of several large hill forms along with distinctly smaller ones. It can be the fall color contrast of brilliant yellow birch or larch drifts adjacent to somber and dark green spruce-fir forest. But quality in vividness does not reside only in conspicuous contrast. It can be subtle, the repetition of softly modulated color or form differences. And, since the landscape is ever changing, vividness may be from the collected impressions of images remembered from spring to fall, the landscape encountered ten miles behind compared to the one at hand. Here are several of the elusive dilemmas of evaluating landscape through aesthetic judgment (Figure 8,5).

Apart from the professional designer-planner evaluating landscape with aesthetic criteria (or design criteria derived from aesthetics and perception), there is the evaluation based upon preferences voiced by the lay public. This is now more in the realm of academic research than a well developed tool for landscape planning. It does not seem feasible for there to be massive testing of public response for every landscape evaluation problem that arises. This would be cumbersome, expensive, and it does not enjoy political support. It is desirable to discover correlations (and departures) between professionally identified evaluation criteria and the values expressed by a participatory public. This mutually beneficial kind of procedure is represented by workshops being conducted by the United States National Park Service concerning Yosemite Valley planning alternatives.[22] It is also found in the psychological research in which the scenic elements of the Connecticut River Valley landscape are the subjects of perceptual response and evaluation.[23]

Qualitative and quantitative assessments need comment. Qualitative evaluations are derived from criteria which are not reducible to simple or precise numbers. Comparative evaluations among and between different visual units are entirely practical, but they are still qualitative. It is also entirely practical to make many measurements of landscape elements—such as relative relief, number and distribution of lakes, or mosaic unit areas of different vegetation types; this can produce some quantification but it does not, cannot put numerical values on the full array of aesthetic criteria. When arbitrary numbers are used to represent certain criteria, they are merely convenient labels to keep track of criteria—summary results remain comparative rather than quantitative.[24] Through the long history of cultural development, aesthetics as the search for beauty has not expressed urgency for the development of a quantification system. Expecting precise measurement of landscape quality is best accepted as an insolvable problem.

Then there is a question, both philosophical and practical, about landscape evaluation. What kinds of decisions are most appropriate in regard to landscape units of different value? Protection and care for high value units is obvious enough but that will serve no purpose if they are made museum pieces within a surrounding landscape where lower visual qualities are marked by neglect. Since high quality landscapes are a rare commodity, it would be a tactical error to spend all the planning-management budget on them. It would make more sense, in protecting overall landscape quality, to put most of the planning and management budget into maintaining the integrity of ordinary landscape which is most apt to be regionally predominant.

A few brief points of emphasis may be made about visual inventories and landscape analysis—including evaluations. They can be and are being used in environmental impact studies and analysis.[25] They have a place in predicting visual impact, and in designing specific changes which are compatible and appropriate to their broader surroundings.[26] Inventories of landscape represent images of specific time; they can be base lines for monitoring change over time; they can serve the legal purpose of demonstrating degradation or loss of values—or the opposite. Visual documentation of the landscape, along with alternative developmental designs, can portray dimensions of change for appraisal by both public and professional audiences. Then landscape analyses have a place in research —in perception, in linkage of ecological-silvicultural techniques to landscape design manipulations, in relating socio-psychological research to problems of landscape planning and resource management. The shopping list of research needs is a long one.

AND FINALLY

The concept of the landscape—wild or otherwise—as a scenic resource with quality is a part of environmental quality. Overall integrity of the visual landscape is at the heart of maintaining aesthetic quality. This applies to both local and regional contexts. It does not mean that sensitive and useful man-made changes are undesirable or degrading—they can be excellent. But quality goes out the window if a policy is adopted to *preserve the best, forget the rest.* If visual-physical integrity of the landscape can be maintained, it leaves options for the future. Generations beyond ourselves deserve a full array of opportunities which are vested in a rich and varied landscape.

FIGURE 1,5   The ordinary or typical landscape of regional foothills. White Cloud Mountains, Idaho. *Author Photo* (page 106)

FIGURE 2,5   Striking Silhouette of dominating landforms makes a feature landscape. Stratified Wilderness, Shoshone National Forest, Wyoming. *Author Photo* (page 107)

FIGURE 3,5   Enclosed landscape with clearly defined floor and walls with definite patterns. Leavitt Meadow, Toiyabe National Forest, California. *Author Photo* (page 108)

FIGURE 4,5   Visual distinction of a three part vegetative pattern, enhanced by fall display of aspen. Carabou National Forest, Idaho. (U.S. Forest Service photo). *Author Photo* (page 109)

FIGURE 5,5   Fast white water — dominant feature in a granite landscape. Tuolumne Cascade, Yosemite National Park, California. *Author Photo* (page 110)

FIGURE 6,5   Water, flat landforms and flat-topped vegetation in a placid, panoramic landscape of vast extent. Yukon River near Ft. Yukon, Alaska. *U.S. Forest Service Photo* (page 111)

FIGURE 7,5   Different from these visual impacts of roads and power corridors, land use patterns can be fitted to landscape, Whiskeytown National Recreation Area, California. *Author Photo* (page 112)

FIGURE 8,5   Landscape of high scenic quality due to unity of enclosure, vividness of dike features, variety in vegetation pattern. Bighorn National Forest, Wyoming. *U.S. Forest Service photo* (page 113)

REFERENCES

1. LITTON, R.B. Jr., *Forest Landscape Description and Inventories.* Berkeley: U.S.D.A., Forest Service, Pacific Southwest Forest and Range Experiment Station, Research Paper PSW-49, 1968; and U.S.D.A., Forest Service, *National Forest Landscape Management,* Vol. 2. Washington: Supt. of Documents, 1974.

2. SANTAYANA, G., *The Sense of Beauty.* New York: Dover, 1955, p. 83.

3. LITTON, R.B. Jr., *et. al., Water and Landscape, An Aesthetic Overview of the Role of Water in the Landscape.* New York: Water Information Center, Inc., 1974, pp. 5-10.

4. LITTON, *op. cit.*

5. LITTON, R.B. Jr., "Aesthetic Dimensions of the Landscape", in KRUTILLA, J., (ed.), *Natural Environments.* Baltimore: John Hopkins Press for Resources for the Future, Inc., 1972, pp. 266-75.

6. JONES, G. and JONES, I., *An Inventory and Evaluation of the Environment, Aesthetic, and Recreational Resources of the Upper Susitna River, Alaska.* Seattle: Jones & Jones for United States Army Corps of Engineers, 1975, p. 80.

7. LITTON, *op. cit.,* (reference 1), pp. 46-58.

8. U.S.D.A. Forest Service, *op. cit.,* pp. 4-15.

9. TETLOW, R.J. and SHEPPARD, S., *Visual Resources of the Northeast Coal Study Area.* Victoria: British Columbia, Ministry of the Environment, Technical Supplement to the Northeast Coal Study Preliminary Environmental Report, 1977, pp. 20-21, 24-27, A.15-A.17; and LITTON, *op. cit.,* (reference 1).

10. LITTON, *op. cit.,* (reference 5).

11. LITTON, *op. cit.,* (reference 1).

12. LITTON, R.B. Jr., "Esthetic Resources of the Lodgepole Pine Forest", in *Proc. Management of Lodgepole Pine Ecosystems Symposium,* Vol. I. Pullman, Washington: Washington State University Extension Service, 1975, pp. 285-296.

13. LITTON, *et. al.*, *op. cit.*, pp. 18-95.

14. LEOPOLD, L.B. and WOLMAN, M.G., *River Channel Patterns: Braided, Meandering, and Straight*. Washington, D.C.: United States Geological Survey, Professional Paper 282-B, 1957.

15. LITTON, R.B. Jr., "Visual Vulnerability", *Journal of Forestry*, 72, 1974, pp. 392-397.

16. TETLOW and SHEPPARD, *op. cit.*, pp. 20-21.

17. *Ibid.*, pp. 9-13; and LITTON, R.B. Jr., "Visual Landscape Units of the Lake Tahoe Region", in *Scenic Analyses of the Lake Tahoe Region*. South Lake Tahoe, California: Lake Tahoe Regional Planning Agency and United States Forest Service, 1971.

18. BEARDSLEY, M., *Aesthetics: Problems in the Philosophy of Criticism*. New York: Harcourt, Brace and World, 1958, pp. 466-470.

19. PEPPER, S.C., *Aesthetic Quality*. New York: Scribners Sons, 1973, p. 44.

20. LITTON, *op. cit.*, (reference 5), pp. 282-287.

21. *Ibid.*, pp. 26-28.

22. U.S. National Park Service, *Yosemite — Summary of the Draft General Management Plan*. Yosemite: National Park Service, 1978.

23. ZUBE, E.H., PITT, D.G. and ANDERSON, T.W., *Perception and Measurement of Scenic Resources in the Southern Connecticut River Valley*. Amherst: University of Massachusetts, Institute for Man and His Environment, Pub. R-74-1, 1974.

24. TETLOW and SHEPPARD, *op. cit.*, pp. A.7-A.14.

25. JONES and JONES, *op. cit.*

26. LITTON, *op. cit.* (reference 15).

PLATE 7 Visual Metaphor: The prairie landscape as a sea of grass.   *B. Sadler Photo* ▶

6 PAINTING, PLACE AND IDENTITY: A PRAIRIE VIEW

Ronald Rees
University of Saskatchewan

INTRODUCTION

A persistent theme in prairie writing is the search for home and identity. The preoccupation reflects, on the one hand, the needs of a species strongly endowed with the home-making instinct, and, on the other, the extent and anonymity of prairie space. The refrain "O bury me not on the lone prairie," however banal, was a cry from the heart. Fear of prairie space, and the accompanying desire to dispel or diminish it, are threads that run through the pioneer letters, diaries and reminiscences. A Welsh home-steader in Saskatchewan, who eventually returned to his Cardiganshire home, left this recollection of his painful first encounter with the prairie: "We had visualised a green country with hills around, and happy people as neighbours — no doubt a naive outlook on so drastic a venture, but one common to many people emigrating at that time. There was something so impersonal about this prairie, something that shattered any hope of feeling attached to it, or ever building a home on it." And of his own quarter section: "What is there to make it different from the rest of the land we've come through . . .? Nothing. Nothing at all."[1]

For those immigrants intent on settling permanently, the problem, therefore, was how to make a home in, and of, a land so lacking in the attributes of uniqueness, familiarity and shelter usually associated with the home place. Although the prairie was particularly intractable, the process of domestication in Western Canada was no different, and no less difficult, than in other pioneer regions. To be home, a new land must first be made to sustain and shelter; then, in Northrop Frye's phrase, it must be imaginatively digested or absorbed.[2] The instruments of domestication are tools in the one case, symbols and images in the other. With them impersonal space is converted into familiar place, or, as an eminent writer and artist puts it, natural environment into human landscape, the latter defined as "a segment of nature fathomed by us and made our home."[3]

117

In new surroundings, the most fundamental level of image-making is naming and mapping. Familiar names for features and places, and symbolic organization of space, make the world seem less intimidating and more comprehensible. But unless they are descriptive and evocative, maps are an objective ordering of space; in Martin Buber's terms they "orient" but they do not "realize."[4] Realization, which is a sympathetic rather than a controlling relationship with environment, can be achieved only through the exercise of imagination. For aboriginal peoples, the avenues are religion, myth and magic; for Europeans, literature and the arts. "An Art," remarked Lawren Harris "must grow and flower in the land before the country will be a real home for its people."[5]

THE PROBLEM OF THE PRAIRIES

The anxiety caused by tension between a mind and surroundings not imaginatively integrated with it is the subject of an autobiographical essay by the historian W.L. Morton. Morton grew up on a farm in Manitoba at the beginning of the century and, as a boy, was disoriented by the conflict between the landscape he saw daily and the landscape in his mind's eye formed by reading Victorian English literature. He writes:

> Landscapes . . . are not only material. There are also the landscapes of the mind . . . These are formed, not by the play of the eye on meadow and forest, valley and mountain, but by books, pictures, television, music. They are cultural landscapes, formed by what the mind, not the seeing eye alone, takes in. As I grew familiar with the actual world in which I chanced to be born, I began, once I learned to read, to enter another and, as it happened, quite a different landscape indeed . . . Thus my actual landscape, the one the neighbours had made and worked in . . . and my literary landscape . . . were in conflict. I had no single vision for both, but had to refocus like one passing from dark to light . . . The difficulty was to reconcile a landscape actually seen and realistically experienced with an internal landscape formed by reading.[6]

Although the experience was not an "existentialist imbroglio," Morton, nevertheless, had a strong desire to see his everyday surroundings transformed through art into a "humane landscape of heightened tone and enriched associations." He desired to see it "not as it could be seen, but as it might be seen."

In the absence of a regional art and literature, Morton could satisfy the desire only by creating his own images. This he eventually did through historical writing. He believes that good history is holistic; it should not only be true to fact but should possess its own integrity, "the truth of total vision." In writing about the West, Morton admits that he was participating in the ancient game of possessing by naming and describing. But his naming was no mere taxonomy. "No country," he writes, "can really be owned except under familiar name or satisfying phrase. To be apprehended by the mind and [made] personal, it requires not only the worn comfort of a used tool or broken-in shoe; it requires also assimilation to the mind, ear, eye, and tongue by accepted, and acceptable, description in word, or line, or colour." For Morton, reconciling the actual with the mind's landscape was a "rite of reassurance fusing the thing seen and the person seeing." The reconciliation of the inner with the outer vision, he concludes, is part of the magic of art and also of the historian's craft.

Morton's anxiety must have been an experience common to all thoughtful settlers. My wife, a native of Saskatchewan, says that throughout her youth she was disturbed by the fact that she did not live in a place or a landscape. These were elsewhere. Like Morton, she could not associate the world she actually lived in with the one acquired through reading. Consequently her surroundings, virtually untouched by poetry, painting, myth, or legend, scarcely entered her imaginative life. There were no oak trees, meadows, or orchards, and at the bottom of a prairie garden, not even the hope of finding little people. The result of the dissociation, as in Morton's experience, was mild distress — the anxiety of placelessness.

At the time of Morton's childhood, and even of my wife's, the imaginative absorption of the prairie had scarcely begun. In pictorial art, such landscapes as existed for the most part had only a token connection with the region. Reasons are not hard to find. For painters born and trained in Europe the prairie was an extremely difficult subject. Vegetation, light, and colour were unfamiliar and the flatness was daunting. As a droll Manitoban remarked to a visiting writer-artist, Edward Roper, "there's a good deal more scenery wanted in this country, ain't there."[7] Arthur Lismer, of the Group of Seven declared that the only interesting objects on the prairie were telephone poles.[8] Even prairie-born landscape painters were dismayed by their subject: Illingworth Kerr, who has spent a lifetime painting the prairie, describes the open parts of it as "the least inspiring environment for a painter."[9]

119

THE EARLY RESPONSES

The response of the first resident painters to so difficult a subject was to concentrate on the few paintable parts of it. Consequently, valleys and wooded slopes became popular subjects. When painters did venture into the open prairie they usually resolved the problems presented by the flatness and the space by raising their horizon lines and filling their foregrounds with figures and objects. Only occasionally, then, do the early paintings convey a sense of the awesome dimensions of the landscape and the distinctive qualities of its light and colour. Filtered through the dark tones and heavy shadows of conventional European painting, the prairie acquired a decidedly old world aura: "the blight of misty Holland, mellow England, and the veiled sunlight of France," as a young Canadian painter remarked at the time (see Figures 1,6 and 2,6).[10]

Yet in spite of their European look, the landscapes seem to have been universally accepted as true statements about the new country. There are several possible reasons for this. Paintings that looked European provided a link with the culture of the homeland and, at the same time, sustained the illusion that the new land was not radically different from the old. In pioneer societies art is an analgesic. Homesick settlers had no desire to be shown the harsh realities of their adopted land. When life itself is simple and demanding, realism, as Morton suggests, can neither inspire nor inform.[11]

As well as comforting by suggesting resemblances between the adopted country and the old, the early landscapes also quieted by offering, in such forms as woods, valleys, and solid, rooted farmsteads, symbols of shelter. For many, the prairie was a landscape of frightening exposure offering few refuges from such physical hazards and discomforts as prairie fire, blizzards, wind, and strong sunlight. In a flat, dry, exposed landscape, valleys assume particular importance, especially when incised. Wallace Stegner, recalling his boyhood experience of the treeless plains of southwest Saskatchewan, writes:

> In a jumpy and insecure childhood . . . sanctuary matters. That sunken bottom sheltered from the total sky and untrammeled wind was my hibernating ground, my place of snugness, and in a country often blistered and crisp, green became the colour of safety. When I feel the need to return to the womb, this is the place toward which my well-conditioned unconscious turns like an old horse heading for the barn.[12]

So strong was the attraction of the valleys that a few settlers left home-steads on the open prairie to live in them. One couple gave up farming on the Regina Plain, for ranching in the Willow Bunch area, on aesthetic rather than economic grounds.[13] Ranches were usually situated in picturesque coulees where there were trees, water, and shelter. Scottish painter James Henderson, equally disturbed by the austerity of the Regina Plain, quickly left it for the Qu'Appelle Valley where he built a small cottage and painted landscapes with such redolent titles as "Sunshine and Shower," "Autumn Hillsides" and "Summer in the Valley."

The need for comfort and security also explains the affection for trails and paths and their frequent appearance in the early paintings. Not only were they symbols of occupance and possession — "an insistence," Stegner writes, ". . . that we had a right to be in sight on the prairie [and] that we owned and controlled a piece of it"[14] — but they provided relief from the inorganic regularity of the rectangular survey system. "The trail," said Hamlin Garland, "is poetry; a wagon road is prose; the railroad is arithmetic."[15] The symbolic importance of trails and paths is also the subject of a poignant poem by a Ukrainian settler:

I found no path, no trail
But only bush and water
Wherever I looked I saw
Not a native (land) — but foreign.
I found no path, no trail
Only green bush
Wherever I looked I saw
A foreign country.[16]

In serving as buffers between the settlers and a difficult environment, the early landscape paintings can be assigned to anthropologist John Bennett's category of "symbolic adjustments" to the prairies. Symbolic adjustments are practices aimed at preserving the image of the old environment in the new. As such, they are not adjustments in the literal sense but, to use Bennett's phrase, "manifestations of resistance to change," a refusal to "give in," along certain symbolic gradients, to the limitations of the habitat.[17]

For some, the resistance began with the selection of a location that offered reminders of the home environment, or idealized memories of it; parkland, for example, was generally preferred to treeless prairie not only because it could provide timber and fuel but because it looked more like home. As Edward Roper put it, "the green plains, the patches of water, the park-like clumps of trees, gave the land a home-like, if not a picturesque look."[18] Sometimes the desire for a picturesque setting outweighed practi-cal considerations. A.F. Kenderdine, one of the first resident painters in the

West moved his house from its original quarter-section to one on a wooded hill nearby that overlooked a lake. "It didn't seem to matter," his daughter wrote later, "that the hill . . . took up most of the 160 acres, leaving a mere fringe to be cultivated. We all loved the place."[19]

As well as choosing settings that reminded them of home, the settlers also undertook physical changes to create the illusion of a sylvan country-side. Trees were planted, streams dammed to make ponds and lakes, and lawns and gardens laid. Many of the changes were functional but they had, nevertheless, an underlying psychological purpose — to create the feeling of home in the subhumid plain. Shelter-belts, as novelist Frederick Niven noted, were a case in point:

> [Shelter-belts] create a sanctuary from the eternal vision of the plains for those who find it monotonous. Within the outer circle of wind-breaks farmers are even now planting fruit trees. You step from the great billiard-board, the expanse of rectangular fields and long straight roads, into a grove where the birds are singing . . . into an oasis where the garden paths twist for a change, and there are green lawns and the billiard-board at once seems remote.[20]

Extensive remodelling of the environment was possible only in such collective enterprises as the building of public institutions. One of Bennett's examples is the Legislative Building in Regina. Towering and high-ceilinged in the 19th century monumental style, it makes no concessions to a flat landscape or to a cold, windy climate. And as if to add insult to an already slighted environment it is set in an artificial landscape of lake, lawns, shrubs and trees that requires unremitting care. There is, Bennett notes, hardly a trace of plains or semi-arid features — boulder landscapes, ravines, native wild flowers, or even aspen clumps. For Bennett, the Legislative Building represents a missed opportunity to promote native forms, but for a C.B.C. broadcaster, closer to the public sentiment, it is seen as a small victory in a battle with an inhospitable environment:

> "This park, this Wascana Lake within it, and indeed the whole tree-studded City of Regina, is a tribute to the sweeter side of man . . . Out here . . . everything must be planted and it will be forever a tribute to Farmer John Saskatchewan that he wanted more than prairie about him — he wanted the softness of nature, the sweet smell of flowers and the comfort of trees. So, in his Capitol he planted them.[21]

CONFRONTING THE OPEN PRAIRIE

Symbolic re-creations of the old country, in both the landscape and in art, are now regarded as expressions of the colonial phase of the development of young societies. Though necessary to uprooted migrants as an emotional bridge between the old country and the new, they could not support indefinitely. To change the metaphor, and continue Northrop Frye's, they were a soft diet that assuaged the hunger for home without satisfying it. To be home, the hard, new land had to be emotionally absorbed. In other words, the "incubus" of the open prairie had to be confronted. For prairie painters of the twenties and thirties, stirred by nationalist and regionalist feelings, the challenge was to replace images of an unassimilated land, Frye's "predigested picturesque," with something more sustaining.

By the twenties, two developments in art had made the task seem formidable. First, Impressionism. To become truly regionalist Canadian painters had to cast off the bonds of academic studio painting. The more adventurous discovered that impressionism—an empirical approach relying on direct sensory experience in the field—was well-suited to painting the unfamiliar field of a new land. A young Toronto painter with strong Western affinities, Charles W. Jefferys, was quick to see its possibilities. The practice of open-air painting in oil using juxtaposed patches of primary colour gave, he noted, "a more brilliant representation of light and atmosphere" and seemed "peculiarly adapted to the expression of high-keyed luminosity and the sharp clear air of mid-Canada."[22] On numerous Western excursions between 1907 and 1924 Jefferys painted a series of watercolours and oils, evoking the brilliance of light and the delicate, shimmering quality of vegetation, that are now widely regarded as the first truly regional paintings of the prairies (see Figures 3,6 and 4,6). Jefferys was the antithesis of the studio-bound academic painter, describing his method of direct and close observation as "sketching a la Ruskin."[23] A remark made at the time to a young Saskatchewan painter suggests the freshness of his approach to the prairies: "Wolf willow, lovely stuff to paint!"[24]

Impressionist techniques helped to resolve the problems of painting prairie light and colour, but held no answer to the question of how to deal with the flatness and the space. Although the naturalism of his results set him apart, Jefferys, like most of the immigrant painters, was a painter of valleys and smiling summer landscapes. The need for a painter able to express the totality of the landscape is evident in an observation-cum-appeal made by the heroine of the best known of the early realist novels, Sinclair Ross's *As For Me And My House*. "Only a great artist," says Mrs. Bentley, "could ever paint the prairie, the vacancy and stillness of it, the bare essentials of landscape, sky and earth."[25] A Saskatchewan painter, Illingworth Kerr, if not the "great artist" Mrs. Bentley had in mind, was first to show the way. Kerr studied in Toronto with members of the Group of Seven and, touched by the Group's crusading nationalism, came back to the West determined to paint the prairie through native rather than European eyes. But on his return, in the late 1920's, he discovered that his methods were no match for his subject. Strong colours and bold brush strokes—the Group's stock-in-trade—are suited to a solid landscape of rock, lakes, and trees, but not to one as insubstantial as the prairie. "For the visual artist," Kerr notes, "the prairie is a great dome of sky, frequently cloudless and without much colour, and a flat plain of earth receding to the horizon. If one sits on the ground it all disappears behind the nearest weeds or grass."[26] Recognizing that a literal approach would defeat the painter, he concluded that the only way to paint a landscape of "vast scale" and "power of mood rather than tangible form," is to distill one's vision of it into simplified, symbolic statements. These distillations he called idea or archetypal landscapes, and, as assemblages of characteristic features rather than actual places, they resemble the geographer's notion of landscape (see Figure 5,6). By setting the painter free from the particular experience, the symbolic or abstract approach encourages a more searching relationship with the environment. Painters who adopted it were able to paint not so much what they saw but, as Frye puts it, what they experienced in their seeing.[27]

The approaches initiated by Jefferys and Kerr, one focussing upon details in the landscape and the other evoking its totality, have become standard among prairie landscape painters (see Figures 6,6 to 11,6). They are consonant with our ways of seeing the prairie. In a landscape where nothing interrupts the view, the eye leaps from one's near surroundings to the far distance. Painters, therefore, must choose between microcosm and macrocosm. The choice, and its effects upon our vision were noted by Jefferys:

> Landscape that has no striking topographical shapes, that consists of earth, sky, space, light, air, reduced to their simplest elements and baldest features. In this severe austerity the grasses, the flowers, the shrubs, claim our attention, attract the eye and assert their individual charms. Vision becomes subtly discriminating, compares hues, tones, colours, all of them within a narrow range of what the artist calls values; yet under this compulsory and refined analysis, revealing an astonishing variety . . . .[28]

Painters who, like Jefferys, opted for the parts rather than the whole have revealed a landscape of delicate, subtle forms. Visually and ecologically the prairie is a landscape of scarcity rather than plenty that appeals only to still, quiet observers. One suspects that motorized tourists on mountain pilgrimages never appreciate it. These see only the macrocosm of immense plain that can make even hardened residents, to borrow again from Northrop Frye, "hide under the bedclothes." Fredelle Bruser Maynard, drawing upon childhood memory, describes its uncompromising austerity in a passage that could serve as commentary for almost any of the paintings of the region's first popular "macrocosmic" painter (see Figure 9,6):

> There were no "prospects" on the prairies — only one prospect, the absolute, uncompromising monotony of those two parallel infinities, earth and sky. I draw it in my mind's eye, with a ruler — road and tree, farmhouse and elevator, all spare and simple and hard-edged, with a line of telephone poles slicing the distance. Movement in this landscape has no more consequence than the leap of a jackrabbit across a dusty road. The stillness is the reality.[29]

THE DEGREE OF SUCCESS

In spite of the acclaim accorded both gentle microcosmic and sublime macrocosmic images, our behaviour suggests that we have not fully accepted either. Although lived in, the prairie is not yet home. In planning and architecture, for example, there is little evidence of appreciation of the landscape's Euclidean simplicities and acceptance of its equally uncompromising climate. In an environment that suggests a need for simple, streamlined forms and protective enclosures, gables catch the wind and broad, straight, unstopped streets draw eyes, longing for restraint, outward to the prairie. There is also little tangible evidence to suggest a general willingness to reject the picturesque vision and the norms of humid lands (see Figure 12,6). An acceptable tree is as likely to be a spruce as an aspen, and an appropriate surround a demanding lawn rather than native cover. In new suburbs and shopping centres sylvan names proliferate: Forest Grove and Wildwood are two recent Saskatoon examples. The prairie is also unheralded by the popular and commercial art of the region. Calendar pictures are more likely to feature mountains and pastoral lowlands than prairie, while the cover of the current telephone directory of — as it likes to pride itself — *the* prairie province shows a stream flowing through a sunlit forest glade. Admittedly, the photograph was taken in northern Saskatchewan but the scene is hardly typical of even the boreal region.

These observations are of more than academic interest since the desire to efface dryland realities can result in heavy public investment in questionable undertakings. John Bennett's most telling example of a symbolic adjustment is the South Saskatchewan dam. Built ostensibly for irrigation, against the advice of a Royal Commission, it is now destined chiefly for power development and recreation. Official emphasis on the recreational and aesthetic potential of the scheme, essentially an afterthought, has increased at the same rate as doubt as to its practicality.

Reluctance to accept the more difficult aspects of the prairie environment is understandable and needs no apology. When reality is hard to bear the natural reaction is to ignore or disguise it — in effect, to hide under the bedclothes. Despite official claims to the contrary, for much of the year the prairie is inhospitable and uninteresting, and in all seasons the proportions — the "outrageous flats" of one poet[30] — are in-human. Such a landscape, like Hardy's Egdon Heath, can appeal only to instincts "subtler and scarcer" than those satisfied with the sort of beauty called charming or fair. This being so, the prairie may always be a rare, or acquired, taste. But if appreciation of it is to be more widespread, then it seems that we must look to a more thorough permeation of the artist's vision. Lord Clark at least is hopeful, asserting that however long it might take this vision is always accepted unconsciously by the uninterested man.[31]

The impact that a powerful image can make upon a blinkered sensibility was made clear to me during a showing of the film Dr. Zhivago one winter night in Saskatoon. In it there is a scene, filmed, I believe, in Alberta, showing a dark train snaking its way across a great snow-covered plain. It was such a chilling image that it caused a girl sitting nearby to shudder involuntarily and remark to her companion: "Oh, I'd hate to live there." A few moments later she began to laugh with embarrassment. The laughter, I like to think, was the beginning of her seeing.

FIGURE 1,6   A.F. Kenderdine — "Country Scene" n.d.
Mendel Art Gallery, Saskatoon (p. 127)

FIGURE 2,6   A.F. Kenderdine — "Whitcomb Trail" n.d.
Glenbow -Alberta Institute (p. 128)

FIGURE 3,6   C.W. Jefferys — "Saskatchewan River at Battleford" 1924 or 1927
Private Collection (p. 129)

FIGURE 4,6   C.W. Jefferys — "A Prairie Trail" 1912 Art Gallery of Ontario (p. 130)

FIGURE 5,6   Illingworth Kerr — "Straw Stacks, March Thaw" 1935
Glenbow -Alberta Institute (p. 131)

FIGURE 6,6   Reta Cowley — "Uncultivated Land and Distant Farmstead" 1976
Private Collection (p. 132)

FIGURE 7,6   George Jenkins — "Prairie Pothole" n.d. Private Collection (p. 133)

FIGURE 8,6   George Jenkins — "The Old Toal Place" n.d. Private Collection (p. 134)

FIGURE 9,6   Robert Hurley — "Prairie Landscape" 1953 Private Collection (p. 135)

FIGURE 10,6   Illingworth Kerr — "Prairie Landscape" n.d. Private Collection (p. 136)

FIGURE 11,6   William Kurelek — "How Often at Night" 1972
Mendel Art Gallery, Saskatoon (p. 137)

FIGURE 12,6   Cover of a Commercial Box — The Persistent Picturesque (p. 138)

128

COLOR SKETCH OF NEAR SATELLITE, 1967

HURLEY

REFERENCES

1. DAVIES, E. and VAUGHAN, A., *Beyond the Old Bone Trail*. London: Cassell, 1960, p. 35.

2. FRYE, N., "Canadian and Colonial Painting", in *The Bush Garden*. Toronto: Anansi, 1971, pp. 199-202.

3. KEPES, G., *The New Landscape*. Chicago: Paul Theobold, 1956, p. 18.

4. BUBER, M., *Daniel*. New York: Holt, Rinehart and Winston, 1964, pp. 61-81.

5. Quoted in MELLEN, P., *The Group of Seven*. Toronto: McClelland and Stewart, 1970, p. 111.

6. MORTON, W.L., "Seeing an Unliterary Landscape", in *Mosaic*, 3 1970, pp. 1-10.

7. ROPER, E., *By Track and Trail Through Canada*. London: W.H. Allen, 1891, p. 51.

8. KERR, I., "Painting, Personality and The Prairies". Unpublished manuscript. Kerr was taught by Lismer.

9. *Ibid.*

10. Quoted in STACEY, R., *Charles William Jefferys 1869-1951*. Kingston: Agnes Etherington Art Centre, 1976, p. 13. The observation was made about Canadian landscape painting as a whole.

11. MORTON, W.L., *Manitoba: A History*. Toronto: University of Toronto Press, 1957, p. 418.

12. STEGNER, W., *Wolf Willow*. New York: Viking, 1966, p. 22.

13. HAMILTON, M.A. and HAMILTON, Z.M., *These are the Prairies.* Regina: School Aids and Text Book Pub. Co., 1948.

14. STEGNER, *op. cit.*, p. 271.

15. Quoted in KELLS, E., "Pioneer Interviews". Unpublished manuscript, c. 1935, p. 173. Glenbow Alberta Foundation, Calgary.

16. Quoted in KOSTASH, M., *All of Baba's Children.* Edmonton: Hurtig, 1977, p. 17.

17. BENNETT, J., "Habitat, Institutions and Economic Development in Saskatchewan". Unpublished manuscript, 1965, pp. 24-27. Saskatchewan Archives, Saskatoon. See also BENNETT, J., *Northern Plainsmen.* Chicago: Aldine, 1969, pp. 85-88.

18. ROPER, *op. cit.*, p. 102.

19. BEAMISH, M., Biographical notes on file, u.d. Glenbow Alberta Institute, Calgary.

20. NIVEN, F., *Canada West.* Toronto: J.M. Dent, 1930, p. 48.

21. FISHER, J. W., "Dust but never Despair", C.B.C. Radio Broadcast, March 19, 1944.

22. Quoted in STACEY, *op. cit.*, p.14.

23. *Ibid.*, p. 19.

24. KERR, I., "Notes for a lecture on Prairie Landscape Painting", Unpublished manuscript.)

25. ROSS, S., *As For Me and My House.* Toronto: McClelland and Stewart, 1957, p. 59.

26. KERR, I., *op. cit.*, (reference 8).

27. FRYE, N., "Lawren Harris: An Introduction", in *The Bush Garden.* Toronto: Anansi, 1971, pp. 207-212.

28. Quoted in STACEY, *op. cit.*, p. 67.

29. MAYNARD, F.B., *Raisins and Almonds*. Toronto: Doubleday, 1972, p. 187.

30. STEVENS, P., "Prairie: Time and Place", in *Nothing but Spoons*. Montreal: Delta Canada, 1969.

31. CLARK, K., *Landscape into Art*. London: John Murray, 1949, p. 47.

PLATE 8   A Sense of Time in Place: Badland coulees dramatically expose the geological history of the plains landscape.   *B. Sadler Photo* ▶

7 THE VISUAL QUALITY OF THE NATURAL ENVIRONMENT IN PRAIRIE FICTION

Dick Harrison
University of Alberta

INTRODUCTION

> "Every man sees in nature that which he brings eyes to see."
>
> Bertram Tennyson, Moosomin, N.W.T., 1896.[1]

I was tempted to call this paper "The Prairie Seen Through Ink," because my main object is to explain why fiction writers cannot be depended upon for faithful verbal pictures of the prairie. One reason is obvious: after suffering all the limitations to perception which he shares with the visual artist, the writer must pass his vision of the environment through a non-visual medium. By the same token, he is likely to be under the sway of fashions in both visual and literary arts, while his depiction is further complicated by tensions between the verbal and the visual. In descriptions of the prairie environment these complicating factors are all apparent in forms conditioned by those more local problems peculiar to a literature which must assimilate a strange, un-European landscape to an essentially European culture.

VERBAL BLINDNESS

To uncover the cultural roots of the problem, it may be necessary to go back to pre-literary prairie experience. While it is widely acknowledged that the prairie makes a powerful visual impression, early explorers and travellers were surprisingly reticent about the prairie landscape, though they could occasionally be quite eloquent about the more familiar parkland scenery to the north. This failure to describe the prairie brings to mind what Professor Ronald Rees has said in his excellent article "The Scenery Cult" about early responses to mountain scenery in eighteenth-century Europe.

143

Because mountains had long been eschewed as objects of contemplation, considered "monstrous excrescences of nature," the first attempts at description did not progess far beyond such phrases as "delightful Horror." Rees goes on to say, "In part, the artificiality of the response was due to the inadequacy of the existing vocabulary to deal with the new sensations."[2] Similarly, the prairie, another of nature's excesses, had no close counterpart in the experience or the language of early British and Canadian travellers. But there seems to be more to the problem. After his tour of the West in 1848, Canadian artist Paul Kane painted the prairie scene near Saint Boniface to look suspiciously like a European countryside. Kane, of course, had been raised in Upper Canada and trained in the salons of Europe. But when in his *Wanderings of an Artist* he subsequently describes the Saint Boniface area as resembling "the cultivated farms of the old country," one begins to suspect more than a deficiency of vocabulary or techniques.[3] It seems likely that both verbal limitations and artistic conventions conspired with his settled habits of perception to make it difficult for Kane to see the totally new environment with any clarity, and the same could probably be said for most of his contemporaries.[4]

The writings of some other early travellers illustrate clearly that the effects of language on visual perception went far beyond deficiencies in vocabulary. Verbal conventions and literary fashions exerted an active force on the selection and shaping of landscape description. Alexander MacKenzie, in his *Voyages* dating from 1789 and 1793, provides virtually no description of the prairie, yet when he reaches the park belt he describes a point on the Clearwater River at some length:

> This precipice, which rises upwards of a thousand feet
> above the plain beneath it, commands a most extensive,
> romantic, and ravishing prospect . . . a most beautiful
> intermixture of wood and lawn . . . stately forests, re-
> lieved by promontories of the finest verdure.[5]

From the terms MacKenzie uses, "romantic," "ravishing," "wood," "lawn," "stately forests," terms which might have come from an eighteenth-century topographical poem, it is clear that the scene lends itself to familiar conventions of landscape description. It is therefore easier to see, respond to, and record. In the light of this description, MacKenzie's failure to describe the prairies can be seen, in part, as a form of "verbal blindness" induced by the literary conventions of his day.

His condition can also be related to artistic conventions. I would suggest, in light of Professor Rees's article, that MacKenzie was specifically addicted to the "picturesque," and that the reticence of earlier travellers may also have owed something to the conventions of landscape art in their day.

If in seventeenth-century art "Mountains and other phenomena suggesting powers in nature beyond human control such as oceans, storms, and winter . . . were avoided,"[6] then the earliest travellers such as Henry Kelsey (1690) and Anthony Henday (1754) could be forgiven for failing to see that the prairie constituted a landscape.

It was not, in fact, until the Romantic Movement licenced an interest in mountains, oceans, deserts and such extremes of nature that extensive descriptions of the prairie began to appear. William Francis Butler, the most celebrated and romantic of the early travellers, recorded this impression after his tour of the West in 1870-71:

> The capes and headlands of what once was a vast inland sea now stand away from the shores of Winnipeg. Hundreds of miles from its present limits these great landmarks still look down on an ocean, but it is an ocean of grass. The waters of Winnipeg have retired from their feet, and they are now mountain ridges rising over seas of verdure.[7]

Butler brings a romantic's enthusiasm for mountains and oceans to his appreciation of the prairie, and the comparison with a seascape would have been an effective way of giving form to the unfamiliar topography for his intended British readership. Butler's *The Great Lone Land* was a hugely popular and influential book, and the prairie-sea metaphor, like the phrase he coined for his title, became standard in early prairie descriptions. He caught the imaginations of Canadians as well as British travellers, but as the survival of the poetic term "verdure" suggests, it is difficult to distinguish invention from convention in Butler's descriptions.

His reliance on specifically literary convention is emphasized when we look beyond the landscape to his description of the Cree Indian as another feature of the new environment:

> Poor, poor fellow! his virtues are all his own; crimes he may have, and plenty, but his noble traits spring from no book-learning, from no school-craft, from the preaching of no pulpit; they come from the instinct of good which the Great Spirit has taught him; they are whisperings from that lost world whose glorious shores beyond the mountains of the Setting Sun are the lost dream of his life.[8]

In its general conception the passage is an echo of Rousseau, but for a closer parallel we must go back to Pope's "Essay on Man":

Lo, the poor Indian! whose untutored mind
Sees God in clouds, or hears him in the wind;
His soul, proud Science never taught to stray
Far as the solar walk, or milky way;
Yet simple Nature to his hope has given,
Behind the cloud-topped hill, an humbler heaven;[9]

The pathos, the conception of an unschooled mind which is yet given an intimation of heaven by "Nature" or "the Great Spirit," especially the verbal echoes of the opening lines, "Lo, the poor Indian!" and "Poor, poor fellow!" give the impression of Butler working with Pope's poem open before him or, more likely, indelibly printed in his mind before he encountered the Indians. His actual encounter seems to have contributed nothing further to his description except the possibility of criminal tendencies in the Cree. Similarly, romantic conventions of landscape may have helped the traveller to record impressions of the prairie, but to what extent they helped him to see it with distinctness and particularity is uncertain. The fiction of the Nineteenth Century shows little evidence of perceptions developed beyond the lyric response in Butler's book. With few exceptions, the writers merely exploit the setting as a convenient wilderness against which to plot their old-world adventures.

A HAZE OF RHETORIC

The fiction of the early Twentieth Century reflects a desire, by no means confined to the literary imagination, to see the newly settled Canadian prairie as the garden of the world. One of the most popular writers, Nellie McClung, rarely included a landscape description, but when she did, it reinforced that "garden" view.

> The grain was first beginning to show a slight tingle of gold. It was one of those cloudless sunshining days in the beginning of August, when a faint blue haze lies on the Tiger Hills, and the joy of being alive swells the breast of every living thing. The creek, swollen with the July rain, ran full in its narrow channel, sparkling and swirling over its gravelly bed, and on the green meadow below the house a herd of shorthorns contentedly cropped the tender after-grass.[10]

The scene could have been lifted from any one of a hundred immigration posters issued by the railroads, the land companies or the Government.

The impression of natural plenty, ease and pastoral contentment pervades them all. I am reminded of Henry Nash Smith's comment in *Virgin Land* to the effect that observers who visited Kentucky in the early days tended, as a result of intense promotional campaigns, to see the region "through a haze of rhetoric."[11]

The garden myth, if borrowed from the earlier phase of American history, developed its own peculiarities in the British dominions. Ralph Connor, by far the most popular Canadian writer of the era, reveals further qualities of the garden in a description of a prairie river:

> The firm green sward, cropped short by a succession of campers' horses . . . the flowing river with its soft gurgling undertone, the upstanding walls of the popular bluffs in all the fresh and ample beauty of the early summer drapery, the overarching sky, deep and blue, through which peeped the shy stars, and the air, so sweet and kindly, breathing down about them. It was all so clean, so fresh, so unspoiled to the boy that it seemed as if he had dropped into a new world . . . .[12]

The world is new in the absolute sense; the garden is edenic, regenerative because man has the renewed opportunity for total harmony with nature and therefore with the universe. Prevailing conceptions of the Canadian West at the time conspired with styles in popular landscape art and the literary fashion for sentimental romance to encourage this kind of idyllic portraiture. The edenic implications are sometimes quite explicit, as in a comment by Arthur Stringer's narrator in *Prairie Wife* to the effect that "We're labouring to feed the world, since the world must have bread, and there's something satisfying and uplifting in the mere thought that we can answer to God, in the end, for our lives, no matter how raw and crude they may have been."[13] A more extensive reading of these and other romances of the period reveals that the garden is British, the vision imperial, and that the pastoral imagery is dominated by a hazy identification of the natural order with the divine order and the human order of empire.

PERSPECTIVE EXERCISES

The romances of pioneering typified by the work of Connor, McClung and Stringer were followed from the mid-1920's to the 1950's by a period in the fiction sometimes termed "prairie realism," which was to some extent a reaction against the earlier work. The realists' descriptions of the natural environment reveal some of the effects of that long cultural process of learning to view a strange landscape, but they also reveal more of the difficulties involved in perceiving it and committing it to a verbal form. Two salient features mark these descriptions off from the earlier ones.

The first of these features, as one might expect, is an effort at greater accuracy and precision. Frederick Philip Grove's *Over Prairie Trails* (1922) is in this respect a landmark in prairie descriptive writing. Here Grove struggles with the problem of describing a wind-sculptured snow-drift:

> The interesting parts of the structure consisted in the beetling brow of the cliff and the roof of the cavity underneath. The brow had a honey-combed appearance; the snow had been laid down in layers of varying density . . . and the counter currents that here swept upward in a slanting direction had bitten out the softer layers, leaving a fine network of little ridges . . . .[14]

Grove begins with a few figurative expressions, "beetling brow," "roof of the cavity," "honey-combed," which induce a degree of sensory vividness to his description, but the more exact he attempts to be ("layers of varying density," "counter-currents") the less visual he becomes. Grove's descriptions are sometimes more sensuous or more impressionistic, but they share this tendency to a diagrammatic, abstract precision. The new desire for accuracy does not necessarily improve the visual representation of the prairie.

The second feature to strike the reader is the unrelieved grimness of the later descriptions. The prairie realists generally were convinced that the garden had been a dangerous illusion and that man was actually alienated from the prairie environment in a profound spiritual sense. They were as determined to portray a hostile and threatening nature as the sentimental romancers had been to deliver a chaste and sunlit prairie. The culmination of the trend, as of the whole movement toward prairie realism, is in the work of Sinclair Ross, whose *As For Me and My House* portrays a prairie deteriorated in the 1930's from garden to wilderness to desert. This description of the little town of Horizon is typical:

> The dust clouds behind the town kept darkening and thinning and swaying, a furtive tirelessness about the way they wavered and merged with one another that reminded me of northern lights in winter. It was like a quivering backdrop, before which was about to be enacted some grim, primeval tragedy. The little town cowered close to earth as if to hide itself. The elevators stood up passive, stoical. All around me ran a hurrying little whisper through the grass.[15]

148

Ross's descriptions do not tend toward abstract precision; they are vivid and compelling partly because of their sustained metaphors. Elsewhere Ross describes the prairie in terms of a night sea or a lifeless moonscape. But despite the fact that the prairie desert image of the 1930's has become, for many Canadians, the essential prairie, it was shaped, especially in its tone, by factors other than the drought. Like the earlier garden image, it shows the influence of literary fashions. Ross was undoubtedly influenced by the naturalistic fiction made popular by such writers as Theodore Dreiser, Sinclair Lewis and Frank Norris. Their underlying philosophy of determinism which tended to cast man as a fated victim of his circumstances virtually demanded a stark portrayal of environment.

Beyond literary fashions there remains the persistent demand of the novelists' form that the visual setting be an organic part of the novel's structure. There is a kind of dark animism in Ross's description of Horizon; the setting is instinct with sinister forces. The dust clouds move with a "furtive tirelessness"; the town "cowers," and the whispers in the grass are ominous. The metaphor of a theatre, repeated elsewhere in the novel, is especially significant. Like a stage backdrop, the visible setting is designed to serve the human drama being enacted before it. The hostilities and fears of Ross's narrator, Mrs. Bentley, are clearly projected upon the environment; as she says later, "we think a force or presence" into the stillness and vacancy of the landscape. The visual quality of the prairie must therefore be subordinated to the literary demand that a setting interact organically with the character and actions of a novel.

To risk a wild speculation, I would suggest that both these traits in the realists' descriptions — the projection and the tendency to abstraction — are encouraged by the spare, incomplete visual quality of the plains landscape. A passage from Wallace Stegner's *Wolf Willow* (1955) illustrates what I mean:

> On that monotonous surface with its occasional ship-
> like farm, its atolls of shelter-belt trees, its level ring of
> horizon, there is little to interrupt the eye. Roads run
> straight between parallel lines of fence until they inter-
> sect the circle of the horizon. It is a landscape of circles,
> radii, perspective exercises — a country of geometry.[16]

Here, eighty-five years after Butler's journey, is the familiar prairie-sea metaphor. The farms still seem as impermanent as ships. The entire scene does not include a single human being; the observer, like the characters in so much prairie fiction, is alone, facing the hugeness of simple forms which make up this landscape. His imagination is powerfully stirred yet can find little to seize upon in trying to assimilate the strongly felt impressions. To describe it he is reduced to the abstraction of geometry. Elsewhere Sinclair Ross refers to the prairie as "the bare essentials of a landscape," while W.O. Mitchell terms it "the latest common denominator of nature," which again descends to the level of mathematical abstraction. In Marshall McLuhan's terms, the prairie presents a "cool" image — low definition, high participation — which tempts the mind to impose some pattern of its own and imposes little constraint upon the growth of feelings evoked in the observer. The shift from prairie as garden to prairie as desert is the more easily accomplished.

The most notable exception to the preference for grim description of the prairie environment during the "realist" period — W.O. Mitchell — also had the most sensuous style of the time. Take, for example, the first encounter with the prairie of young Brian O'Connal in *Who Has Seen the Wind*:

> He had seen it often, from the verandah of his uncle's farmhouse, or at the end of a long street, but till now he had never heard it. The hum of telephone wires along the road, the ring of hidden crickets, the stitching sound of grasshoppers, the sudden relief of a meadow lark's song, were deliciously strange to him. Without hesitation he crossed the road and walked out through the hip-deep grass stirring in the steady wind; the grass clung at his legs; haloed foxtails bowed before him; grasshoppers sprang from hidden places in the grass, clicketing ahead of him to disappear, then lift again.
>
> A gopher squeaked questioningly as Brian sat down upon a rock warm to the back of his thighs. He picked a pale blue flax-flower at his feet, stared long at the stripings in its shallow throat, then looked up to see a dragonfly hanging on shimmering wings directly in front of him . . . .
>
> And all about him was the wind now, a pervasive sighing through great emptiness, unhampered by the buildings of the town, warm and living against his face and in his hair.[17]

The description is visually evocative with its "haloed foxtails," "flax-flower" and "shimmering wings" of the dragonfly, but it should be noted that more of its vividness recruits the senses of hearing and touch. In Mitchell's writing there is a balance of the senses which precludes the creation of set verbal "pictures." He has spoken privately about the tyranny of the visual and the dearth of other sense impressions in literary description. In his latest novel, *The Vanishing Point* (1973),[18] perspective exercises become a central symbol. In this novel about the collision of native and white cultures, the perspective exercises to which Stegner has compared the prairie landscape are seen as an artificial ordering device, a paradigm for the imposition of an alien European world-view on the plains and their native peoples. Through various characters in the novel, visual "perspective" is closely related to the specifically imperial moral perspective of the White settlement period apparent in the early romances of pioneering. In this way the tyranny of the the visual sense becomes identified with certain ways of ordering what we see, imposed by fashions in the literary and visual arts which in turn express definite and limited ways of understanding life. Mitchell's later work makes it clear that as the fiction of the prairie becomes more sophisticated, the relationships between the visual and the verbal become more complex and more strained.

INTERNAL LANDSCAPES

Younger writers such as Margaret Laurence, Robert Kroetsch and Rudy Wiebe probably share Mitchell's objection to the imperial perspective; their fiction expresses a strong desire to escape traditional white views of the cultural history of the prairie. While they may express no concern about the tyranny of the visual sense, their fiction is for a variety of reasons not strongly visual. The styles of contemporary fiction generally do not lend themselves to extended landscape description. Further, the environment of which these novelists write, the agrarian West, is essentially passed, absorbed into the new urban-industrial prairie as agribusiness. The visible prairie has therefore become less an object of use and more an object of contemplation; it represents an accessible past in which the writers must seek the significance of their own lives. Where visual descriptions appear, they are less likely to be serving aesthetic ends than figuring forth a complex series of ethical and epistemological questions relating to the search for ancestral roots in the prairie soil.

Take, for example, Hagar Shipley's brief description of the prairie at the beginning of Laurence's *The Stone Angel*. It is contained in an explanation of her father's Scots exile mentality:

> [The highlanders] lived in castles, too, every man
> jack of them, and all were gentlemen. How bitterly
> I regretted that he'd left and had sired us here, the
> bald-headed prairie stretching out west of us with no-

thing to speak of except couch-grass or clans or chit-
tering gophers or the gray-green poplar bluffs, and
the town where no more than half a dozen decent
brick houses stood, the rest being shacks and shanties,
shaky frame and tarpaper, short-lived in the sweltering
summer and the winter that froze the wells and the
blood.[19]

The prairie is glimpsed, briefly but honestly, because the aging Hagar recog-
nizes that her childish acceptance of her father's alienated view was a
betrayal of her real, immediate heritage. The question is moral more than
aesthetic — not how the landscape looked but how she should have looked
upon it. In Margaret Laurence's reflective scenes the visible prairie is less
described than evoked as contributing to a character's state of mind and
feeling. The landscapes are internal.

Robert Kroetsch's prose might be considered a more direct denial of the
tyranny of the visual. In his "post-modern" style of fiction, all outward
appearances are likely to be locked in an ironic tension with the underlying
significance of events. The description of Wildfire Lake as first seen by
Hazard Lepage in *The Studhorse Man* is typical. In the lake Hazard sees a
mustang colt and a Cree Indian:

It was as if they had just bobbed to the surface. Hazard
had looked a minute earlier and there was nothing to be
seen but whitecaps. He was up on a high bank that is
almost a cliff; he was there partly just to look, partly
because he was homesick for the sight of a little water.
The flat parklands break suddenly, and you see a valley
not in front of you, below you. The squares of farmland
are gone and below you a wooded coulee, like the crack
in a fat lady's ass, glides a scar of earth down to a long
narrow lake. The man in the water wasn't riding the
horse; he was swimming beside it with his arm around
its neck, as if one or the other of them was about to
drown.[20]

The description is vigorous and irreverant; the name of the lake is a contradiction in itself, suggesting the ironic role it plays. The two figures in the
lake are inexplicable (the Indian shortly disappears), fateful (the colt draws
Hazard into the errant life of a studhorse man), and suggestive of the legacy
of the West Hazard is inheriting (the native Cree presents Hazard with the

wild horse, symbol of the plains). The questions raised by the scene are more epistemological than aesthetic: is someone — Hazard, the narrator, the reader — intended to learn something about the West from this scene as opposed to orthodox recorded history? Is it possible to learn anything from appearances, since we begin to distort them the moment we attempt to find significance in them?

Rudy Wiebe is probably the most visual of the major contemporary novelists of the Canadian prairie. His two most recent novels, *The Temptations of Big Bear* and *Scorched-Wood People,* because they deal with the Indian and Metis people in the Nineteenth Century, quite naturally demand extensive descriptions of the environment. More importantly, Wiebe has declared his intention to reach back beyond the conceptual frameworks of White history to recover the sensory experience in which he may discover the significance for the present of a prairie past from which he feels we have been systematically cut off. He describes the historical novelist's task of re-creation:

> Through the smoke and darkness and piled up factuality of a hundred years to see a face; to hear, and comprehend, a voice whose verbal language he will never understand; and then to risk himself beyond such seeing, such hearing as he discovers possible, and venture into the finer labyrinths opened by those other senses; touch, learn the texture of leather, of earth; smell, the tinct of sweetgrass and urine; taste, the golden poplar sap or the hot, raw buffalo liver dipped in gall.[21]

In his attempt to unwrite history back to its source in immediate sensory experience, Wiebe evidently regards "the finer labyrinths opened by those other senses" as a more potent resource than the visual sense. After all, we have visual documents and pictures of Big Bear, but how many experiences of sound, touch, taste and smell do we find preserved in our archives? And when writing of the Indian people in particular, Wiebe is striking a different balance of the senses in an effort to recapture what McLuhan would call the tactile world of pre-literate tribal man.

Wiebe's representation of the richness of the tactile, tribal world is part of a larger effort to redress a moral balance between the contending claims of the White and Indian cultures in the prairie past. Even his most visual passages serve the same moral purpose. Here, for example, is the description of Cree Chief Big Bear's view of his last Buffalo hunt:

He lay against the ground completely, and the earth warmth grew in him and slowly each sound and movement and colour and whispy smell of the living world worked in him: one bull pawed dust over himself; an eagle lay upon the blue sky; a wolf studying the calf twitched slightly on his belly stretched behind sage; an ant carried her egg past his nose as a gopher emerged from his burrow, staled, and then quivered erect as he became aware of the long shadow motionless on the ground beside him; grass singing sweetly. Spread there completely alert but empty except of the hunt, the total consuming unconscious joy of one more run merging with mus-toos-wuk given once more to the River People, at last before him again, sixteen only and hardly more than one belly-stretching meal yet everything for life and this moment that could ever be asked.[22]

At least four of the salient features of this passage are typical of Wiebe's descriptions: sensory detail is minute; continuity is impressionistic rather than logical; the prairie environment is part of a rich natural cycle, a living whole in which the Indian takes part; the tone is dominated by a moral concern for the fate of the natural man being over-run by a mechanistic, unnatural civilization.

It is significant that the native peoples have, in the 1970's, reappeared as part of the fictional prairie environment. The writers are searching for a clearer understanding of their region. Some say they are in search of ancestral roots with a local, regional habitation. The search necessarily involves them deeply in portrayal of the natural environment, which is expected to yield up some of the answers to their questions. In one sense of the term, it is a "vision" of the prairie they seek, yet their portrayals are only incidentally visual. As in earlier phases of prairie writing, the expressed visual quality of the prairie environment is subject to literary fashion, but in this immediately contemporary fiction, the tensions between the visual and the verbal are tightened by one further paradox. In seeking a more meaningful version of their own past, the writers are "going Indian," rebelling against the literate, linear, visual perspective of post-Gutenberg European tradition. If there is a tyranny of the visual sense which they must combat in the interests of a tactile, tribal perception of the prairie, it is a tyranny established by the history of their own verbal medium, their own print culture.

CONCLUSION

To understand the deficiencies in verbal renderings of the prairie land-scape in fiction, it is necessary to go beyond the basic problem of capturing a visual effect in a verbal medium. From the earliest to the latest writings, the imposition of an essentially European culture upon an un-European landscape can be identified as a factor, though one which has operated in shifting and unexpected ways. If the early writers were at a loss for visual and literary conventions to give form to the chaos of impressions evoked by a totally unfamiliar environment, contemporary writers are caught in a tension between established conventions of their culture and what they see as the uniqueness of prairie experience. Complex relationships between cultural order and visual form are probably inevitable, since in fiction, as in any art, the visual rendering of a landscape is necessarily an implicit statement about man's relationship to his environment. It is not surprising that the romances of pioneering, expressing that relationship in edenic metaphors, should have tended to sentimental excesses of description. By comparison the stark landscapes of the realists, expressive of man's spiritual alienation from the land, are refreshing but not essentially more devoted to visual fidelity. Contemporary writers, struggling to free them-selves from the limitations of a culture distinguished by a tyranny of the visual sense, offer a visual world which is elusive and often ironic. Thus, while the fiction has come a long way from MacKenzie's "lawns" and "state-ly forests," the visual quality of the natural environment in contemporary novels remains fragmentary, dominated by the other imperatives of the novelist's form and intentions.

REFERENCES

1. TENNYSON, B. *The Land of Napioa*. Moosomin, N.W.T.: Spectator Printing and Publishing Company, 1896, p. iv.

2. REES, R. "The Scenery Cult, Changing Landscape Tastes over Three Centuries," *Landscape* 19 (1975), p.44.

3. KANE, P. *Wanderings of an Artist*. Toronto: Radisson Society, 1925 [1859], p. 49.

4. HARRISON, D. *Unnamed Country, The Struggle for a Canadian Prairie Fiction*. Edmonton: University of Alberta Press, 1977, p. 26. Some of the ideas presented here about the development of prairie fiction appeared in other forms and contexts in that book.

5. MACKENZIE, A. *Voyages*. Toronto: Courier, 1911 [1801], p. cxxxi.

6. REES. *Op. cit.*, p. 42.

7. BUTLER, W.F. *The Great Lone Land*. Edmonton: M.G. Hurtig, 1968 [1872], pp. 138-139.

8. BUTLER. *Op. cit.*, p. 242.

9. POPE, A. *Essay on Man*. Bk. I, 11. 99-104.

10. McCLUNG, N. *Sowing Seeds in Danny*. Toronto: Briggs, 1918 [1911], p. 72.

11. SMITH, H.N. *Virgin Land*. Cambridge: Harvard University Press, 1950, p. 142.

12. GORDON, C.W. *The Foreigner* by Ralph Connor, pseud. New York: George Doran, 1909, p. 246.

13. STRINGER, A. *The Prairie Wife*. New York: A.L. Burt, 1915, pp. 59-60.

14. GROVE, F.P. *Over Prairie Trails*. Toronto: McClelland and Stewart, 1960 [1922], p. 76.

15. ROSS, S. *As For Me and My House*. Toronto: McClelland and Stewart, 1957 [1941], p. 59.

16. STEGNER, W. *Wolf Willow*. New York: Viking, 1955, p. 7.

17. MITCHELL, W.O. *Who Has Seen the Wind*. Toronto: MacMillan, 1947, p. 11.

18. MITCHELL, W.O. *The Vanishing Point*. Toronto: MacMillan, 1973.

19. LAURENCE, M. *The Stone Angel*. Toronto: McClelland and Stewart, 1968 [1964], p. 15.

20. KROETSCH, R. *The Studhorse Man*. Richmond Hill: Pocketbooks, 1971 [1970], pp. 67-68.

21. WIEBE, R. "On the Trail of Big Bear," a paper delivered to the Western Canadian Studies conference in Calgary, March, 1974, pp. 1-2.

22. WIEBE, R. *The Temptations of Big Bear*. Toronto: McClelland and Stewart, 1973, p. 127.

PLATE 9   Stimulus and Symbol: Old Faithful as a spectacular and
sacred national monument.   *B. Sadler Photo* ▶

8 TOWARDS MODELS OF ENVIRONMENTAL APPRECIATION

Allen Carlson
University of Alberta
and
Barry Sadler
University of Victoria

This volume presents a selection of essays in the interpretation of environmental aesthetics. It takes us from scenic wildlands to urban flatscapes, from appraisals of beauty to attacks on ugliness, and from the views of experts to the reactions of everyman. After following these paths the fundamental issues which are being pursued may be profitably restated and reexamined. What constitutes aesthetic quality in environmental contexts? How are we to develop a better appreciation of its significant aspects? Who is to undertake the role of interpretation?

Some concluding thoughts on these questions are offered here. We attempt to make formal distinctions among certain themes and approaches which are illustrated in the preceding essays and implicit in the structure of the field. The analysis leads towards their organisation into alternative models for the appreciation and determination of the aesthetic values of the environment. It is undertaken as much to promote further discussion as to summarize the present one. Our postscript, in other words, is both a concluding note and a next step.

A BASIS FOR DISCRIMINATION

As noted in the introduction, the aesthetic quality of the environment is best understood as an experiential product of the interaction between man and his surroundings. It is used to describe the properties and values of the environment which are appreciated for their intrinsic worth. More often than not, these evoke heightened forms of sensibility and emotion, ones which are out of the ordinary in some measure (even though the context may be familiar). Yet this delineation of the nature of the aesthetic takes us only so far and more explicit definition is needed before further theorising can take place. We begin with a brief summary of the way certain philosophers have approached this question.

The type of experience recorded above is reminiscent of that which the philosopher and educator John Dewey called consummatory. By this expression Dewey wished to distinguish a *creative* and *deliberate* type of experience from one characterised by passive reception of and response to simple sensation.[1] It provides the initial key to a better understanding of environmental aesthetics. For, according to Dewey, consummatory experience is always in part aesthetic, but the aesthetic need not dominate. Our particular interest, therefore, is in that subset of consummatory experience which has traditionally been considered aesthetic and in which the quality of the environment is the sole focus.

Philosphers often characterise aesthetic experience as involving disinterested or purposeless attention.[2] It is undertaken as an end in itself, rather than as a means to some ulterior purpose. Under this concept, the nature of aesthetic experience is defined not in terms of the constraints imposed upon it, but in terms of the lack of constraints which characterise other forms of experience. Such a definition allows us to differentiate, in principle, between aesthetic and non-aesthetic forms of environmental experience.

The basis for discrimination is the limits imposed by interests on the range and character of man's reactions with his surroundings. Experience of environment which is motivated by particular purpose is non-aesthetic. It is no longer in scope nor broader in duration than is necessary. By contrast, aesthetic experience, being undertaken for its own sake, represents a fuller multi-dimensional interaction, one which contains the potential for welding a range of human values and environmental properties.

Understanding aesthetic experience of environment in this way also allows us to draw distinctions among certain concepts which are widely employed in the field, namely aesthetic quality, scenic beauty, and sensory pleasure. These concepts are applied in various ways in the literature, and their meaning is blurred and overlapping. We suggest that the differences among them can be understood and clarified in terms of the extent to which an experience is consummatory and disinterested and in terms of the factors which determine the degree to which it is.

THE DIMENSIONS OF THE AESTHETIC

The full experience of *aesthetic quality* represents a peak phase of environmental appreciation. It is a complete and unified act, which arises when a milieu is absorbed by the senses, conceptualised by the intelligence, and moulded by the imagination. The result is an image rich with expressive and symbolic meaning. Such experiences are typically symbolised in art and articulated in poetry, rather than revealed through everyday response. Only relatively few people, through temperament or training, are able, like John Muir, to "interpret the rocks, learn the language of flood, storm and avalanche—and get as near to the heart of the world as I can."[3] A profound sense of being at one with the world, whether revealed in a blade of grass, the stone of the city, or the vastness of land and sky, is an abiding and universal value. It represents the quintessential form of aesthetic quality which is gained from the interaction of man and environment. To achieve this transcendent state is a rare occurrence because the processes of socialisation appear to work to reduce our capacity for deep aesthetic contact with nature.[4] Wilderness travel, as undertaken by the latter day disciples of Muir, stands as one modern example of an attempt to regain this experience. It is largely motivated by aesthetic values and yields composite feelings of challenge, solitude, beauty, harmony, and diversity.[5]

The experience of *visual or sensory pleasure*, by contrast, is a more interested and less consummatory kind of environmental interaction. It involves the immediate effects on our well being of all we see, hear, and smell, from the scale of trees and signs to that of landscape and townscape.[6] The main focus is on the perceptual rather than the cognitive aspects of environmental experience, emphasising the episodic sequences of visual pleasure and sensory delight which we derive from everyday encounters with our physical surroundings. Such patterns of response can be contrasted with unitary acts of deep appreciation, although this distinction is not easy to sustain in practice. Environmental perception and cognition, after all, are interacting processes which are integrated with the coded structure of past experience.[7] No aesthetic experience thus stands alone but is related to all that has gone before. Sensory pleasure, like deeper aesthetic insight, is a developed response to environmental displays. It is, in theoretical terms, the lowest common denominator of aesthetic response. And it also serves as an important contemporary baseline against which to judge the humaneness of habitats and on which to develop our potentialities for heightened appreciation.

Scenic beauty is popularly considered to be the central dimension of man's aesthetic relationship with environment. It represents the highest common factor of the quality which is derived from our focused acts of environmental appreciation. As such, it may be placed conceptually between aesthetic quality and sensory pleasure. This interpretation is suggested by the philosopher George Santayana who defined beauty as objectified pleasure, i.e., pleasurable response realised in the object of appreciation.[8] In environmental contexts, it is expressed as a particularised response to vis-

161

ible, morphological form—to the natural and cultural landscape as a scene. To view landscape as scenically beautiful is a partial rather than a fully consummatory experience. Only the material aspects of the environment are visually unified while the evocation of beauty represents the dominant expressive meaning. From a theoretical standpoint, it may be contrasted with aesthetic experiences which involve all the senses and the diverse range of symbolic and expressive meanings which intelligence and imagination make possible, and also distinguished from the more transient feelings of pleasure derived from sensory encounters. Yet it is important to remember that beauty and, in particular, natural beauty, evokes aesthetic responses which form an important part of our hierarchy of values and represent a perennial source of self-actualisation for which man strives once the more basic biological and psychological needs are satisfied.[9]

ALTERNATIVE PARADIGMS

The factoring of these dimensions lays the groundwork on which to base formal paradigms of environmental aesthetics. Each concept implies a distinctive orientation towards the physical milieu and focuses on certain modes of perception and evaluation. In this section, we attempt to briefly but explicitly extend these notions of environmental appreciation in order to trace alternative models of structuring experience. These may be termed as the object model, the landscape or scenery model, and the environmental model.

A normal way to conceptualise our surroundings is as a set of *distinct objects*, focusing our attention on those which relate to our purposes. When we conceptualise the environment in this manner the result is often aesthetic appreciation only at the level of sensory pleasure. A particular natural object or cultural artifact is appreciated as a source of sensory pleasure much as we appreciate certain pieces of sculpture. This typically involves abstracting the object from its larger context and concentrating on its immediate perceptual features. The pattern and fragrance of a flower, the structure and form of a pebble, even the design and color of a billboard are sources of sensory pleasure when we isolate them as objects of our appreciation. Although this type of appreciation allows for aesthetic pleasure where otherwise we may experience only the mundane objects of our everyday surroundings, its object orientation is clearly a limiting and fragmenting way of conceptualising our total environment. In general, it seems to be the least apt model for the aesthetic appreciation of an environment.

More characteristic of contemporary approaches to environmental aesthetics is to conceptualise the landscape as *scenery*.[10] The cityscape from the balcony and the wildscape from the viewpoint are paradigm sources of scenic beauty. To consider the environment as landscape or scenery broadens our view; particular objects become part of a larger design, which is visually unified and expressive of beauty. But the view remains limited. We appreciate only what we *see* and then appreciate it as something distinct from us. The landscape when the object of scenic appreciation is always other than our immediate surroundings; it is a prospect which we look upon from the outside, removed in regard to both distance and involvement. Moreover, if the view is chaotic, unyielding to visual unification, it may fail to become the source of any aesthetic satisfaction. Thus, the scenery model, though a widely favoured conceptualisation, nonetheless imposes limitations upon aesthetic experience.

To conceptualise the environment as *a total setting* opens the door to an appreciation of its full aesthetic quality. Rather than an object or scene which is distinct from the appreciator, this model yields a context for the appreciator, one of which he is a part. Appreciation is not a matter of looking at the features of things, but of living with them.[11] Whether the environment is a neighborhood or a wilderness area, its reality rather than its appearance becomes the focus of appreciation. Since the observer is a part of that which he appreciates, both its nature and his nature are constitutive of its aesthetic quality. Health and vitality in an environment so conceptualised are a source of beauty, disintegration and decay a source of ugliness. This orientation forges in the natural environment a connection between aesthetic quality and ecological soundness and in the man-made environment between aesthetic quality and human well-being.[12]

SOME IMPLICATIONS FOR RESEARCH AND PRACTICE

The previous discussion yields a means to place in perspective certain important themes in environmental aesthetics. It has particular implications for the discussion among students of the field on the role that expert opinion and public preference should play in the determination of aesthetic values. And this, in turn, leads to a consideration of the general issue of how to strive to maintain or improve the aesthetic quality of the environment.

Visual or sensory pleasure, as noted previously, constitutes a baseline for an aesthetically acceptable environment. Given the continued decline in visual values, a focus on the elimination of ugliness is often the only practical course of action. Because it requires only an object orientation and demands less on the part of the observer, the assessment of sensory pleasure in environmental features is properly the role of everyman. We need no experts to tell us when these look, sound, or smell disturbing. Yet as the

163

rewards of sensory pleasure are comparatively unspectacular and the result of its lack subtle and insidious, the key role in our lives is all too easy to overlook. Environments which are low in aesthetic quality or scenic beauty are frequently neglected both in theory and in practice. The result is that they often cease to be a source of even simple pleasures. Public opinion is rightly offended when the environments we live in become ugly by omission.

The possibility of appreciating beauty, by contrast, requires somewhat more both of the environment and of the observer. Built or natural environments considered to be beautiful are the exception rather than the rule, and to develop a genuine intimation of this quality requires that observers are educated to perceive somewhat more than the casual eye may notice. Often this education of taste cannot be achieved without the aid of the expert, although in many cases it is implicit in our cultural and artistic heritage. We need to understand the role of prevailing norms and standards and the way these have shifted over time. The scenic orientation is so ingrained as a contemporary paradigm that the appreciation of beauty typically overshadows both of the other two dimensions of aesthetic experience. As a result, not only is the need for achieving a minimum of sensory pleasure from our environment overlooked, but also the possibility of experiencing its total aesthetic quality is not pursued.

Experiencing the aesthetic quality of an environment requires much of the appreciator, but paradoxically less of the environment. An environment need not necessarily have design nor even be lacking in ugliness; it need only be such that it can be fully experienced consummatorily and disinterestedly. Perhaps it is any environment in which we can transcend ourselves and of which we can feel a proper part, i.e., any setting with which we can achieve an environmental orientation. Such an achievement requires of the observer a heightened sensitivity and awareness, both to environments in general, and to the particular context to which he responds. He must not only experience its character, but understand it and his role within it. Consequently, various kinds of expert opinion are useful in aiding us to discover and appreciate aesthetic quality in our environments.

Environmental appreciation ranges on a spectrum from sensory pleasure to aesthetic quality. The determination of values on this scale presupposes conceptualising the environment in a certain way. Much current work in the field involves an object or a scenery orientation. This emphasis is evident in some of the essays in this volume. It is important because sensory pleasure and scenic beauty are major dimensions of the aesthetic. Studies on these themes, in particular, have strong practical implications. Landscape evaluation, whether at a macro or micro scale or undertaken through the analysis of visual features or social response, yields information that has more or less direct utility for decision-making in environmental conservation and design. The importance of continuing in these directions and linking the results more strongly to planning requirements hardly requires further underlining.

On the other hand, the conditions necessary for a fuller appreciation of the aesthetic quality of the environment remains relatively neglected. Such an orientation, in our view, constitutes an important direction for further research in the field. The education of sensitivity to what the environment is and what it might become will not reverse immediately the contemporary decline in aesthetic values. But it should pay dividends over the long term. For a change in our paradigm of the environment ultimately changes the way we alter and shape our surroundings. The study of this relationship has been a long standing concern for geographers and other students of man-environment relationships. It thus offers a range of possibilities for interdisciplinary collaboration to yield holistic perspectives on the kinds of environments which can enrich our lives. We hope that the collection of essays gathered together in this volume constitutes a step in this direction.

REFERENCES

1. See DEWEY, J. *Art as Experience* (New York: Capricorn, 1958), especially Chapter III. Dewey's view is indicated by the following quote: "Experience occurs continuously, because the interaction of live creature and environing conditions is involved in the very process of living. Under conditions of resistance and conflict, aspects and elements of the self and the world that are implicated in this interaction qualify experience with emotions and ideas so that conscious intent emerges. Often times, however, the experience had is inchoate. Things are experienced but not in such a way that they are composed into *an* experience . . . . In contrast with such experience, we have *an* experience when the material experienced runs its course to fulfillment. Then and then only is it integrated within and demarcated in the general stream of experience from other experiences . . . . Such an experience is a whole and carries with it its own individualizing quality and self-sufficiency. It is *an* experience." (p. 35).

2. This tradition runs from the Third Earl of Shaftesbury (1671-1813) through Immanuel Kant and Edmund Burke to the present time. For an historical survey, see Jerome Stolnitz, "On the Origins of 'Aesthetic Disinterestedness'", *Journal of Aesthetics and Art Criticism*, 20 (1961), pp. 131-143. Stolnitz provides a summary of the summary of the basic elements of this type of view in *Aesthetics and the Philosophy of Art Criticism* (Boston: Houghton Mifflin, 1960), pp. 32-42. He defines "aesthetic experience" as the total experience had with "disinterested and sympathetic attention to and contemplation of any object of awareness whatever, for its own sake alone."

3. Quoted in BACHELDER, L., ed. *Nature Thoughts*, Peter Pauper Press, N.Y., 1965, p. 23.

4. DREWS, E.M. and LIPSON, L. *Values and Humanity*, St. Martin's Press, N.Y., 1971, p. 36.

5. SHAFER, E.L. and MIETZ, J. "Aesthetic and Emotional Experiences Rate High with Northwest Wilderness Hikers", *Environment and Behaviour*, Vol. 1, 1969, pp. 187-197; and GARDNER, J. "The Meaning of Wilderness: A Problem of Definition", *Contact*, Vol. 10, No. 1, 1978, pp. 7-33.

6. LYNCH, K., *Managing the Sense of a Region*, The MIT Press, Cambridge, Mass., 1976.

7. See NEISSER, U., *Cognition and Reality*, W.H. Freeman, San Francisco, 1976.

8. SANTAYANA, G., *The Sense of Beauty*, New York: Collier, 1961, Part I, especially pp. 40-45.

9. MASLOW, A., *Towards a Psychology of Being* (2nd edition), Van Nostrand, New York, 1968.

10. The central place of the scenery orientation in current work is quite obvious. This can be appreciated by noting the titles of some well known articles in the field; for example, "Not Every Prospect Pleases: What is Our Criterion for Scenic Beauty?", "Landscape Aesthetics: How to Quantify the Scenics of a River Valley", "The Assessment of Scenery as a Natural Resource", "Mapping the Scenic Beauty of Forest Landscapes", and so forth.

11. This orientation is implicit in many of Aldo Leopold's remarks; for example: "In country, as in people, a plain exterior often conceals hidden riches, to perceive which requires much living in and with". See especially LEOPOLD, A., "A Taste for Country" in *A Sand Country Almanac with Essays on Conservation from Round River* (New York: Ballantine, 1974), pp. 177-233.

12. For a discussion relevant to the preceding three paragraphs, see CARLSON, A., "Appreciation and the Natural Environment", *Journal of Aesthetics and Art Criticism*, 37 (1979), pp. 267-275.

PLATE 10 The Pleasures of the Harbour. *I. Norie Photo* ▶

THE CONTRIBUTORS

Jay Appleton is Professor of Geography, University of Hull, U.K., author of *The Experience of Landscape* and chairman of the Landscape Research Group.

Allen Carlson is Associate Professor of Philosophy, University of Alberta, Edmonton.

Richard Harrison is Associate Professor of English, University of Alberta, Edmonton.

R. Burton Litton is Professor of Landscape Architecture, University of California, Berkeley, and Senior Researcher, U.S. Forest Service.

Douglas Porteous is Professor of Geography, University of Victoria, B.C.

Ronald Rees is Associate Professor of Geography, University of Saskatchewan, Saskatoon.

Ted Relph is Associate Professor of Geography, University of Toronto, Ontario.

Barry Sadler is Consulting Associate to the School of Resource Management, Banff Centre for Continuing Education, Alberta and a part-time member of faculty, Department of Geography, University of Victoria, B.C.